COMBINED HEAT AND POWER

*A Practical Guide to
the Evaluation, Development,
Implementation and Operation of
Cogeneration Schemes*

Robin T. Griffiths

First published in Britain in 1995 by Energy Publications, an imprint of Professional and Corporate Communications, Livanos House, Granhams Road, Great Shelford, Cambridge CB2 5LQ.

The author and publishers have used their best efforts in researching and publishing this work but they do not assume, and hereby disclaim, any liability to any party for any loss or damage caused by errors or by omissions in *Combined Heat and Power: A Practical Guide*, whether such errors or omissions result from negligence, accident or any other cause.

A CIP catalogue record for this book is available at the British Library.

ISBN 1 874334 04 8

Design & Typeset by Flair plan photo-typesetting Limited, Stansted, Essex.

Printed and bound in Great Britain by
Biddles Limited, Guildford and King's Lynn

 This symbol appears within the book beside charts, tables and forms expressly provided for direct photocopying and use by the reader.
   Reproduction of those charts, tables and forms for sale or publication elsewhere is explicitly excluded from this provision.

Dedication

The book is dedicated to Jackie and Holly: Jackie for her tireless work on the manuscript and boundless understanding over the 3 years of research and writing; Holly for giving me the inspiration to keep going on the occasions when my enthusiasm faltered.

Contents

Acknowledgements

The key to producing a practical guide on any topic, in my view, is to talk to as many people as possible who have first hand experience of the subject matter. For this reason, there are many companies and individuals who have assisted me with research for the book who, for practical reasons, cannot all be listed here. Thank you for your help.

I would, nevertheless, like to mention by name the following people who read parts and in some cases all of the manuscript for the book and provided invaluable guidance and help:

Mr. D. Andrews BSc. (Hons.) CEng. FIPlantE. MInstE. MIDGTE. Technical Director, Power Gasifiers International Ltd.

Mr. R.J. Arthur DipEE. CEng. MIEE. MIDGTE. Regional Facilities Engineering Manager, National Westminster Bank plc.

Mr. H.F. Maurer IEng. FIPlantE. FIDGTE. MIMgt. Energy Efficiency Co-ordinating Engineer, Thames Water Utilities.

Mr. K.R.M. Swinburne BSc. (Hons.) CEng. FIHospE. MIMechE. MASHRAE. Partner, Cambridge Consulting Engineers.

Mr. D.D. Ward P. Eng MIMarE. IPF (France) MIDGTE. MASME. MASHRAE, Site Operations Manager, Guy's Hospital, The Guy's and St. Thomas NHS Trust.

In addition, the following organisations have been particularly helpful with the provision of technical data and illustrations:

Banks plc
Brush Electrical Machines Ltd.
Centrax Ltd.
Clayton Thermal Products Ltd.
EA Technology Ltd.
The Energy Technology Support Unit
European Gas Turbines
George Meller Ltd.
Leverton Power Systems
Nedalo (UK) Ltd.
Peter Brotherhood Ltd.
Siemens plc
Select Power Consulting Ltd.
Senior Thermal Engineering Ltd.
Solar Turbines Incorporated

Abbreviations

ASHRAE	–	American Society of Heating, Refrigerating and Air Conditioning Engineers	kWhe	–	kilowatt hour of electrical energy
			kWt	–	kilowatt of heat rate
			LHV	–	Lower Heat Value
AVC	–	Automatic Voltage Control	LTHW	–	Low Temperature Hot Water
AVR	–	Automatic Voltage Regulation	LV	–	Low Voltage
°C	–	degree Celsius	m	–	metre
CCGT	–	Combined Cycle Gas Turbine	mbar	–	millibar
CEM	–	Contract Energy Management	mg	–	milligram
CFC	–	chlorofluorocarbon	MJ	–	megajoule
CHP	–	Combined Heat and Power	MTHW	–	Medium Temperature Hot Water
CO	–	carbon monoxide	MW	–	megawatt
CO_2	–	carbon dioxide	MWe	–	megawatt of electrical power
CT	–	Current Transformer	MWh	–	megawatt hour
dB	–	decibel	MVA	–	megavolt amp
dB(A)	–	A-weighted sound level	MVAr	–	megavolt amp reactive
DCF	–	Discounted Cash Flow	NGC	–	National Grid Company
EGR	–	Exhaust Gas Recirculation	NH^3	–	ammonia
g	–	gram	nm^3	–	cubic metre of gas at normalised conditions of 0°C and 1013 mbar
GJ	–	gigajoule			
HHV	–	Higher Heat Value	NO_x	–	nitrogen oxides
HRC	–	High Rupturing Capacity	NPV	–	Net Present Value
HV	–	High Voltage	OCEF	–	Overcurrent and Earth Fault
HVAC	–	Heating Ventilating and Air Conditioning	PC	–	Personal Computer
			PME	–	Protective Multiple Earthing
Hz	–	Hertz	ppm	–	parts per million
IC	–	Internal Combustion	REC	–	Regional Electricity Company
ISO	–	Internal Standards Organisation	SCR	–	Selective Catalytic Reduction
K	–	Kelvin	SI	–	Systeme International d'Unites
kg	–	kilogram	SO_2	–	sulphur dioxide
kJ	–	kilojoule	STOD	–	Seasonal Time of Day
kPa	–	kilopascal	UBH	–	Unburned Hydrocarbons
kV	–	kilovolt	VDU	–	Visual Display Unit
kW	–	kilowatt	VT	–	Voltage Transformer
kWe	–	kilowatt of electrical power	v/v	–	volume by volume
kWf	–	kilowatt of fuel energy input rate	£	–	Pounds Sterling
kWh	–	kilowatt hour			

Foreword

By Professor Ian Fells, MA PhD FEng FInstE CChem FRSC FIChemE, Department of Chemical and Process Engineering at the University of Newcastle-upon-Tyne.

Ever since the steam engine was invented and the industrial revolution got under way man has tried to improve its efficiency. When Carnot, and later Clausius, enunciated the laws of thermodynamics our understanding of how heat engines worked was transformed. Heat, obtained by burning coal, gas, wood or whatever, can be used via a heat engine to produce work but some heat is inevitably lost in the process and often wasted. The principle of combined heat and power enables a heat engine-based system to be optimised in terms of energy efficiency so that the maximum use can be made of the chemical energy released by combustion from the primary fuel. That energy is converted into electricity or motion or both, as in the case of a car, a good example of a CHP scheme, and also into lower-grade but still useful heat. It is essential to understand the thermodynamics of the process, but just as important are the economics, so that saving energy also results in saving money.

At last the realisation is dawning amongst politicians and the general populace alike that we must not only husband our resources of fossil fuel but use them as efficiently as possible to reduce the production of carbon dioxide gas which threatens, through the greenhouse effect, to destabilise the weather machine and profoundly change climate throughout the world.

The widespread and increased use of combined heat and power systems will materially improve the efficiency with which we use our fuel. Robin Griffiths' book is an essential source of information on how to put CHP into practice.

INTRODUCTION

Don't skip this section, it's worth reading!

Above all this book is meant to be well written, concise and relevant. The idea is that you, the reader, can get at the information you need as quickly and as painlessly as possible. With this in mind, the book has been broken down into eight chapters to take the topic logically, step by step from the thermodynamic principles right through to the operation and maintenance of a combined heat and power plant.

Although I very much hope that "Combined Heat and Power: A Practical Guide" will be of interest to the academic world, *the book has been written for the practitioner.*

If you are considering combined heat and power and want to know: whether it's worth it; about the technology; how to go about it etc., then this book is for you.

Many publications of this type shy away from engineering detail in the belief that, in this way, they will attract the widest readership. Unfortunately, this often leads to the production of light weight books that skate on the surface of a subject without really tackling the funda-

mentals. It is my firm belief that, expressed in the right way, these important fundamentals are accessible to anyone with a modicum of engineering knowledge. I make no apologies, therefore, for plunging straight into thermodynamics in Chapter 1.

To ensure that the text remains straightforward and relevant to the typical cogeneration practitioner, however, I have employed 'insert panels' for 'non-core topics'. Whilst these panels will, I believe, be of interest to many, their omission in the first reading of a chapter may be of assistance to those for whom much of the chapter content is new.

I believe it is worth pointing out that this book contains no advertising and has not received any financial support from the combined heat and power industry. I have, therefore, been absolutely free to say exactly what I like on all aspects of the technology and have done so. I hope that you will find this frankness both refreshing and illuminating.

1

THE FUNDAMENTALS OF POWER AND HEAT GENERATION

The successful application of any technology demands a basic understanding of the principles behind that technology. This chapter sets out, in plain English, the thermodynamic fundamentals behind combined heat and power.

THE FUNDAMENTALS OF POWER AND HEAT GENERATION

1.1 Heat and Power Concepts

I would like to start by posing the question: "Does combined heat and power (CHP) have some fundamental advantage or is it simply the latest energy efficiency fad?" To examine this question we must first consider the basic ideas of work and heat.

Well everybody knows what 'work' is and what 'heat' is, or do they?

To a thermodynamicist heat and work have specific meanings. Heat is the term used to refer to the transfer of energy between two objects as a result of a temperature difference between the objects. Heat is not, thus, a property of an object like temperature, rather heat describes the movement of energy due to temperature difference.

The term work is analogous to heat in that it is also used to refer to the movement of energy. In the case of work, however, the transfer of energy arises as a result of a force being applied to move an object over a distance. Thus, work is the movement of energy due to physical displacement.

As both heat and work are used to refer to the transfer of energy, they are measured in the same units, namely units of energy. A consideration of 3 practical observations will, nevertheless, demonstrate that the two are quite different.

Observation 1   If a heavy box is moved from A to B by pushing it across a horizontal floor, the box will come to rest, with 100% of the work being done on the box being converted into heat through friction. It can, therefore, be deduced that work can be transformed into heat with 100% efficiency.

Observation 2   Conversely, it is a practical observation that no device, be it steam engine, gas turbine, petrol engine or diesel engine, can transform 100% of the heat input to it into work. It can, therefore, be deduced that heat cannot be transformed into work with 100% efficiency.

Observation 3   Finally, consider a typical swimming pool containing 3,000m³ of water at 30°C. The energy content of the pool water with reference to 15°C is equivalent to the chemical energy content of 5,000 litres of diesel oil i.e. ap-

proximately 50,000 kWh. It is a practical observation, however, that whereas the heat that could be transferred from the pool water cannot, in practice, be used to produce any work, the heat released by burning the 5,000 litres of diesel oil could be used to drive a car two times around the world. It can, thus, be deduced that there is a qualitative aspect to potential sources of heat in addition to the obvious quantitative aspect.

From these practical observations, I would like to draw two conclusions that are pivotal to the economics of CHP:

(a) Work is of more value than heat.
(b) An energy source that can be used to produce work is of more value than one that can produce only heat.

The relevance of these conclusions will become clear in Section 1.3. But where does 'power' come into all this? Well one slightly confusing aspect of the term 'combined heat and power' is that heat refers to a quantity of transferred energy, which in SI units is measured in 'Joules' (J). Power, on the other hand, concerns rate of energy transfer and hence is measured in 'Joules per second' (J/s), commonly known as 'Watts' (W). In fact, power is work done per second.

1.2 The Fundamentals of Heat Engines

Enter the 'heat engine'. Any device which takes heat and converts it into work is known as a heat engine. A detailed discussion of the thermodynamics of heat engines would not be appropriate here. Reference to a text book such as "Engineering Thermodynamics, Work and Heat Transfer" by Rogers and Mayhew (ref. 1) would, however, remind us of the two basic laws of thermodynamics. These laws when applied to heat engines can be stated as:

First Law   "To produce work a quantity of heat must be supplied to an engine."

Second Law   "To produce work not only must a quantity of heat be supplied to an engine but a proportion of that heat has to be rejected, at a lower temperature than that at which it was supplied."

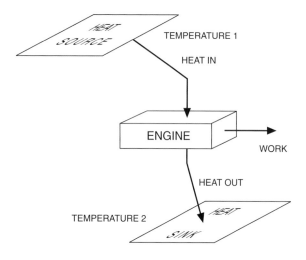

Notes

1. *Carnot Efficiency* = $\dfrac{Temperature\ 1 - Temperature\ 2}{Temperature\ 1}$

where
temperature is absolute temperature in Kelvin (Celsius + 273.15)

Figure 1.2(a)   *The Heat Engine*

The consequence of the first law is that energy is always conserved (except where nuclear reactions are concerned). The consequence of the second law, however, is that a heat engine can never be 100% efficient in the conversion of heat into work. This is a fundamental and inescapable fact. No matter what technological advancements occur, *heat engines can never be 100% efficient*.

The operation of a heat engine in accordance with the second law of thermodynamics is illustrated in figure 1.2(a). It can be shown that the theoretical maximum efficiency of all heat engines is determined by the difference between the temperature at which input heat is available and the temperature at which heat can be rejected. In figure 1.2(a) this is the difference between temperature 1 and temperature 2. The greater this difference the greater the potential efficiency of the engine. Thermodynamicists define the maximum possible theoretical efficiency of a heat engine as the 'Carnot Efficiency'. The expression for Carnot Efficiency is given in figure 1.2(a).

In practical engines, the input heat temperature is limited by materials technology – engine components must be able to withstand the high temperatures. Heat reject temperatures are generally ambient. With current materials technology, the best that heat engines can achieve in practice is the conversion of just over half of the heat input into work i.e. a heat to work efficiency of approximately 53%.

Now let me summarise these points. Work is of more value than heat. To convert heat into work a substantial proportion of the energy input to an engine must be lost – this is thermodynamic law. The greater the difference between the temperatures at which heat is input to and rejected from the engine, the greater the theoretical maximum heat to work efficiency of the engine.

1.3 The Pros and Cons of Combining Power with Heat Generation

In our fossil fuel dominated economies, coal, oil and gas are burned to produce high temperature heat for two purposes. Firstly, fuel is combusted simply to produce high temperatures for space and process heating. The thermodynamic short-sightedness of this is discussed in the insert panel of this chapter. Secondly, fuel is used in engines to generate work and hence power.

'Combined heat and power' or 'cogeneration', as the names suggest, simply combines these two activities. If, to produce power, heat has to be rejected, why not make use of that rejected heat thereby maximising the benefit gained from the fuel burned? Combining the generation of heat and power is the elegant way to utilise our finite fossil fuel resources.

The impact that CHP can have on the consumption of fossil fuels is illustrated by the two diagrams shown in figure 1.3(a).

There are, however, two potential difficulties that arise when heat and power generation are combined. One obvious, the other less so. Firstly, if an engine is to be used to produce both power and heat then demands for both power and heat must be found. Now, power can be transmitted long distances at relatively low cost in the form of electricity. Heat is not so easy. Heat has to be transmitted in the form of hot water or steam in pipe networks at relatively high cost. To gain the potential benefits of cogeneration, therefore, there must be a demand for heat relatively local to the engine plant.

The second potential difficulty concerns temperature. For heat to be useful for space or process heating, it must be generated at the required temperatures. These temperatures will, of course, be higher than ambient. Returning back to figure 1.2(a), we will be reminded that the efficiency of power generation is determined by the difference in temperature at which heat is input to and rejected from the engine. By raising the temperature at which heat is rejected, to make it suitable for space or process heating, the efficiency of power generation is lowered.

To illustrate this point with some numbers, consider the two engines shown in figure 1.3(b). Engine 1 rejects heat at ambient temperature

(20°C) and has a theoretical maximum efficiency of 66%. Engine 2, however, rejects heat at a temperature suitable for space heating (80°C) but has a theoretical maximum efficiency of only 60%. In this case, the price for recovering heat from the engine at a useful temperature is a 9% loss in power output from the plant.

From the above it will be realised that combining heat with power generation entails a trade off between power efficiency and useful heat recovery. As work is of greater value than heat, a relatively high proportion of the heat rejected by an engine must be recovered if the associated loss of power generation efficiency is to be economically justified.

Conventional Use of Fossil Fuels

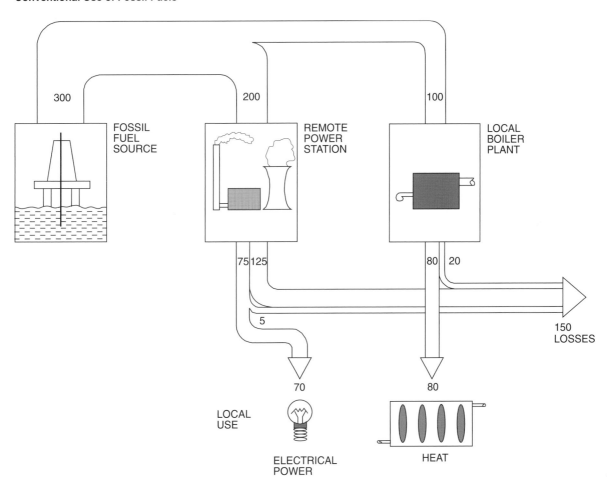

Result: 300 units of fossil fuel energy required to produce 70 units of electricity and 80 units of heat for local use.

Figure 1.3(a)   *The Impact of CHP on the Consumption of Fossil Fuels*

Cogeneration Use of Fossil Fuels

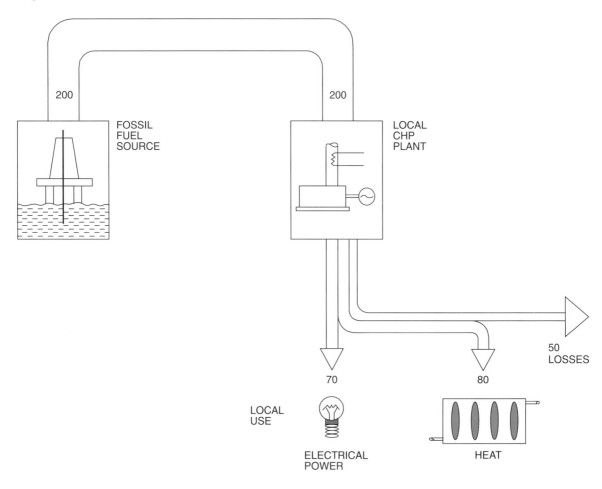

Result: 200 units of fossil fuel energy required to produce 70 units of electricity and 80 units of heat for local use.

Figure 1.3(a) continued   *The Impact of CHP on the Consumption of Fossil Fuels*

THE THERMODYNAMIC ELEGANCE OF COGENERATION

A brief review of the relevant thermodynamics has been undertaken in sections 1.1 and 1.2 of this chapter and two fundamental points have been made: Firstly, work is of more value than heat. Then secondly, the efficiency with which work is delivered by an engine is dependent upon the difference between the temperatures at which heat is input to and rejected from the engine.

It will be concluded, therefore, that the maximum temperature which can be produced from a particular energy source will be a key factor in determining the amount of work that can be generated from that source. This potential of an energy source to undertake work is termed 'availability'. Availability is quantified with reference to ambient temperature using the following expression:

$$A = Q(T_1 - T_2)/T_1$$

where:

A = Availability
Q = Energy content of heat source
T_1 = Temperature produced by heat source
T_2 = Ambient temperature

It can be seen that the expression for availability is simply derived from the formula for 'Carnot Efficiency' given in figure 1.2(a). Sources of heat that can produce high temperatures and therefore have a high availability, are known as fuels.

We are fortunate that on this Planet we have been blessed with adequate supplies of fuels such as oil, gas, coal and uranium. Without them the Industrial Revolution simply wouldn't have happened. These supplies are, how-ever, *finite*. Once burned, their capacity to undertake work is lost for all time. Unlike energy, there is no 'conservation of availability law'. For this reason, these high grade sources of energy must be treasured and consumed with the utmost thought and care.

From the above, it will be realised that burning oil, gas or coal to generate heat simply for space or process heating squanders the remarkable potential of these fuels to undertake work.

The elegance of combined heat and power from a purely thermodynamic perspective, should now be clear. The concentrated capability that fossil fuels possess to undertake work should, whenever practical, be exploited by burning these fuels to generate power first and using the rejected heat to satisfy any space or process heating demands.

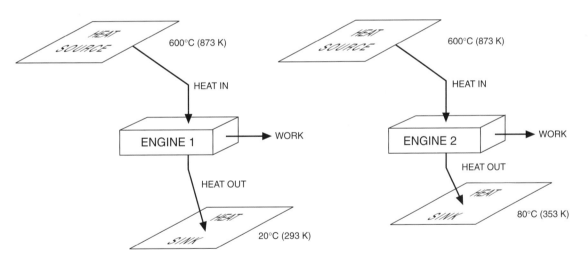

Notes

1. *Carnot Efficiency for Engine 1* = $\dfrac{873 - 293}{873} \times 100 = 66\%$

2. *Carnot Efficiency for Engine 2* = $\dfrac{873 - 353}{873} \times 100 = 60\%$

Figure 1.3(b)   *The Effect of Raising Heat Rejection Temperature*

2

WHEN SHOULD COMBINED HEAT AND POWER BE CONSIDERED?

The cogeneration approach to heat and power production is not appropriate for all sites. As a detailed CHP feasibility study can be expensive, it is important that the likely return on investment in cogeneration is confirmed before substantial funds are committed to any in-depth investigations. This chapter provides guidance on the economic forces that work for and against cogeneration and provides assistance with the preliminary assessment of the likely financial returns from investment in CHP.

WHEN SHOULD COMBINED HEAT AND POWER BE CONSIDERED?

2.1 Economic Factors

Let me start with a statement of the obvious. "To be considered a financial success, a CHP scheme must cut heating and electricity costs to a level where the financial savings achieved provide an acceptable return on the capital investment made". An in-depth examination of this statement will clarify the issues concerned.

The conventional approach to providing heat and electricity to a site is to purchase fossil fuels for use in local boilers to generate heat and to purchase electricity generated centrally by others. In the alternative combined heat and power approach, fossil fuels are purchased and used in an engine to generate both heat and electricity locally. The fundamental economic factors that work to the advantage of each approach are summarised in table 2.1(a).

As a percentage of overall operating costs, electricity transmission costs are relatively small. Essentially, therefore, the four significant economic advantages of central electricity generation have to be outweighed by the more effective use of input fuel, if a CHP scheme is to cut site energy costs. The level of effectiveness required will, of course, depend on a number of factors. From the above analysis, however, it will be realised that for CHP to be an alternative worth considering, there must be a significant demand for both electricity and heat at the site. In addition, as the potential to store either heat or electricity is strictly limited, it follows that demand for electricity must occur at the same time as demand for heat.

The unavoidable conclusion is, thus, that investment in combined heat and power is only likely to prove cost effective at a site that has significant and simultaneous demand for electricity and heat for a high number of hours in the year.

2.2 Local Demand Patterns

In this section, the characteristics of site heat and electricity demands are analysed and the demand patterns for sites where CHP may be applicable are explored.

2.2.1 Heat to Power Ratio

In some publications on CHP, the ratio of heat demand to electricity demand is viewed as pivotal to the consideration of cogeneration at a site:

"If your heat to power ratio is X you should be considering a gas turbine, if its Y then a reciprocating internal combustion engine would probably be appropriate and if its Z then CHP is unlikely to be cost effective."

There are, however, two reasons why this over-emphasis of the importance of the heat to power ratio can be misleading. Firstly, at most sites the ratio of heat demand to power demand is not a fundamental constant but rather it varies from day to night, from weekday to weekend day and from season to season. So on what basis is this magical ratio calculated – on annual averages, weekday averages or daytime averages? Secondly, the cogeneration system at a site does

Table 2.1(a)   *Fundamental Economic Factors*

Factors working to the advantage of local heat generation and the purchase of centrally generated electricity.	*Factors working to the advantage of local heat and electricity generation.*
❏ The use of low cost fuels at large power stations.	❏ The more effective use of input fuels achieved by CHP plant.
❏ Economies of scale achieved in the operation and maintenance of large power stations.	❏ The elimination of electricity transmission losses.
❏ Access to low cost capital finance.	
❏ Government subsidies in the finance and operation of large power stations.	

1

MANAGEMENT TIME AND THE FINANCIAL PERFORMANCE GAP

In the perfect organisation with unlimited finance for capital projects, all schemes with a positive net present value over their lifetime would be implemented. In most real organisations, however, capital is a scarce resource. CHP schemes have, therefore, to compete for finance against a range of other capital projects.

In the 1970's and 80's, expansion through diversification became fashionable. In the 1990's, however, both Government and commercial organisations are shifting decisively back towards concentrating on core activities and contracting out peripheral tasks.

For this reason, consideration of energy projects, including combined heat and power, is seen by many business managers as a diversion of management time which could more profitably be applied to the core business of the organisation. The results of this management strategy can be two-fold. In the first instance, it may become increasingly difficult to get senior management to give serious attention to energy projects. Then, secondly, the financial returns demanded from investment in energy projects may be greater than those demanded from investment in core activities. In other words, a 'financial performance gap' can develop between investment in the core business and investment in energy schemes.

It may be concluded, therefore, that for a CHP project to grab the attention of senior management and then secure the necessary finance, the following will be required:

- The quality of the scheme proposal will have to be exceptionally high in both content and presentation.
- The financial returns generated by the scheme will have to be significantly greater than those from competing core business schemes.

For these reasons, a whole section of the book in chapter 5 has been dedicated to 'Preparing and Presenting the Proposal'.

There is, however, good news on the horizon in the form of fears over global warming. As improving energy efficiency is seen as the most cost effective means to reduce carbon dioxide emissions and hence combat global warming, it has become an area of activity which now attracts widespread attention. Pressure from governments and consumers is translating this attention into interest and commitment from senior management. Whether the financial performance gap will also be eliminated through this process is not yet known, but the signs are hopeful.

Finally there are, of course, alternatives to obtaining internal finance to fund CHP schemes. These important opportunities are discussed in chapter 6.

---

not have to satisfy the complete demand for either electricity or heat. At a site with an annual average heat to power demand of 3:1 it may be more cost effective to invest in a reciprocating internal combustion engine with boost firing of the exhaust gases rather than in a gas turbine, which might have a more 'appropriate' heat to power output ratio.

As we will see later in this section the convergence of demand profiles is the fundamental factor that determines the suitability of a site for combined heat and power, not the heat to power ratio.

On the other hand, when it comes to the choice of engine technology the temperature at which heat is required is usually the crucial consideration.

2.2.2 Seasonal, Weekly and Daily Demand Patterns
As has been mentioned above, depending upon the site, the demand for heat and electricity will vary from season to season, from day to day and from hour to hour. At manufacturing sites where materials processing is involved, seasonal demand may be relatively constant. Unless

Table 2.2(a) *Site Demand Patterns*

Bin[1]	Description	Number of Weeks[2]	Hours per Week[3]	Average Demand – kW		Consumption[4] – MWh	
				Electricity	Heat	Electricity	Heat
1	Winter – Major use hours						
2	Winter – Out of hours						
Total consumption for Bins 1 and 2							
Total consumption from winter fuel bills[5]							
3	Summer – Major use hours						
4	Summer – Out of hours						
Total consumption for Bins 3 and 4							
Total consumption from summer fuel bills							

Notes

[1] The term 'Bin' simply refers to a period of time. The 'Bin approach' to calculation is discussed in Chapter 4.

[2] The number of weeks in winter and summer is a nominal figure chosen to reflect the fundamental seasonal variation of demand at the site.

[3] The number of hours per week of major use is a nominal figure chosen to reflect the fundamental day/night, weekday/weekend day variation of demand at the site. For example, 'major use hours' for an office building will usually be the hours when occupancy is high and for a production facility will be hours when production is taking place.

[4] Consumption is simply calculated from: number of weeks × hours per week × average demand.

[5] Consumption totals from fuel bills should be approximately equal to calculated consumptions. In the case of heat, the MWh figures from fuel bills must, of course, be multiplied by average heat generation efficiency.

manufacturing takes place 24 hours per day 7 days a week, however, there will be a significant reduction in demands at the weekend and at night.

To get a preliminary understanding of demand variations at a site, average heat and electrical demand for a nominal 26 winter weeks and 26 summer weeks should be analysed under the categories given in table 2.2(a).

The monitoring and metering facilities at the site will often not be sufficient to provide this information. If this is the case, rough estimates should be made based on operating experience and approximate calculations i.e. number of boilers in operation, maximum demand from electricity bills, known electrical equipment operating at night.

The figures compiled for table 2.2(a), will give an indication of how demands for heat and electricity vary from day to night and from season to season. In addition, the variation in the ratio of demands will be revealed, the importance of which is discussed in the following section. In a more practical vein the figures are used as the basis for the evaluation of the potential for CHP set out in section 2.6.

2.2.3 Convergence of Demand Profiles

As stated above, the figures determined for table 2.2(a) will give a rough indication of how the demand for heat converges with the demand for electricity from daytime to night-time/weekend during the winter and the summer.

Section 2.1 touched upon the fact that the technologies available to store either heat or electricity are limited. This means that, with the possible exception of ice storage, it is not cost effective in practice to store either, even for short periods of time. If an engine is to be used to generate both heat and electricity at a site there must, therefore, be a simultaneous demand for both the heat and the electricity produced.

2.2.4 Magnitude of Electricity Demand

The actual size of the electricity demand will be an important consideration in determining which type of engine technology will be most suited to the site. This will be seen from section 2.6, where certain technologies are ruled out if

the demand for electricity is too small.

2.2.5 Managing the Heat to Power Ratio

In section 2.2 so far, the demands for heat and electricity have been discussed as fixed items. This is, however, rarely the case. At most sites, opportunities exist to change a demand for electricity to a demand for heat and vice versa through the use of different plant and machinery. In devising a combined heat and power scheme, the transfer of demand from electricity to heat can often be crucial to the cost effectiveness of the proposal.

For the sake of clarity at this preliminary evaluation stage, this topic will be left until Chapter 4 to be discussed in detail. If, however, the site under consideration has a heat to power ratio of less than 1:1 for a significant number of hours in the year, then reference should be made to section 4.4 at the preliminary evaluation stage, to explore the possibilities of increasing this ratio by managing the demands for heat and power at the site.

Table 2.3(a) *Export of Electricity and Heat – Capital Costs of Distribution versus Energy Transportation Potential*

Form of Energy	Transportation Means	Capital Cost[1] per 100 m – £	Rate of Energy Transport[2] – MW	Value of Energy Transported[3] – £ p.a.	Value of Energy p.a. divided by Capital Cost
Electricity:					
Medium Voltage Electricity (415V)	Electrical Cable (TP&N 240 mm²)	5,000	0.3	100,000	20
High Voltage Electricity (11kV)	Electrical Cable (TP&N 240 mm²)	8,000	7.2	2,540,000	318
Heat:					
Low Temperature Hot Water (80°C)	Steel Pipe (150 mm)	10,000	3.3	290,000	29
Medium Temperature Hot Water (120°C)	Steel Pipe (150 mm)	10,000	5.6	490,000	49
Steam (10 bar abs, dry saturated)	Steel Pipe (150 mm)	10,000	11.0	960,000	96

Notes

[1] Capital cost figures are for the supply and underground installation of 100 m of cable/pipe in plain earth, including trenching and re-instatement.

[2] For hot water and steam, fluid velocities of 3 m/s and 60 m/s respectively have been assumed. For electricity a nominal maximum current of 400A and a power factor of 0.95 have been assumed.

[3] Nominal energy unit values of £0.01/kWh and £0.04/kWh have been assumed for heat and electricity respectively. Operating hours of 8,760 per annum have been assumed.

2.2.6 Summary

In summary, the more constant the demand for heat and power at a site the more likely it is that a CHP scheme will prove cost effective. In cases where demands vary significantly, peaks and troughs in the demand for heat and the demand for power must occur at the same time.

2.3 Export

Up to this point, only the potential use of heat and power at the site in question has been considered. There may, however, be opportunities to export either electricity or heat or both to third parties beyond the site. Whether this is worthwhile will, of course, depend on the payments that are received for these exports of energy and the additional capital cost that is entailed.

2.3.1 Heat

Occasionally, the proximity of other sites may make possible the sale of heat in the form of hot water or steam. As will be seen from table 2.3(a), however, the capital cost of distribution pipework in relation to the value of energy transported is relatively high. With the rate of return on capital required by most organisations, the financing of large pipe networks is unlikely to prove an attractive investment. Under most circumstances, therefore, sites will need to be located within a few hundred metres of the CHP plant for them to be considered as candidates for the export of heat.

In deciding whether to offer a supply of heat to a third party, detailed consideration should be given to the type of contractual commitments that will probably have to be made regarding security of supply. These commitments may necessitate the installation of additional plant to provide the required standby heat generating capacity. In financing the CHP scheme, therefore, the export of heat may not provide the assistance that might be expected.

Finally, the heat exported must, of course, be supplied at a lower price than the alternative at the candidate site. If the candidate site currently generates heat using a low cost fuel such as coal or heavy fuel oil, then the price that can be charged for exported CHP heat may be too low even to finance the required pipework runs, let alone to make a contribution to the funding of the CHP plant.

Using an estimate for the cost of generating heat at a candidate site and the distance to that site, the figures from table 2.3(a) can be used to determine the likely financial benefit of exporting heat. Should the security of service for the supply offered be sufficiently high, the candidate site may no longer have to maintain and operate their own heat generating plant. In this case, it may be possible to charge a higher price per unit of heat supplied whilst still providing considerable financial benefits to the customer.

2.3.2 Power

Power in the form of electricity is a more readily exportable product. The various opportunities that may be open to a generator are illustrated in figure 2.3(a). They are in essence:

- Sale of electricity to a Regional Electricity Company.
- Sale of electricity to a National Grid Company (where one exists) making use of Regional Electricity Company cables.
- Sale of electricity directly to a third party making use of Regional Electricity Company cables.
- Sale of electricity directly to an adjacent third party using 'private' cables laid for the purpose.

Exactly which options are available and which will prove cost-effective will vary markedly from country to country depending upon the structure of the electricity supply industry and the regulatory framework imposed. Advice will, therefore, need to be sought from the relevant regulatory authorities in the country concerned. The following paragraphs have, nevertheless, been included to give the reader some broad guidelines on each of the 4 possible options.

The most straightforward of the options is to sell electricity generated in excess of the site requirement, to the Regional Electricity Company to whom the site is already directly connected. The price paid for supplied electricity will be on the basis of a fixed tariff or, in the case of larger sites, an agreed contract. Contractual security of supply obligations will not, in general, be too onerous but equally the prices paid per unit of

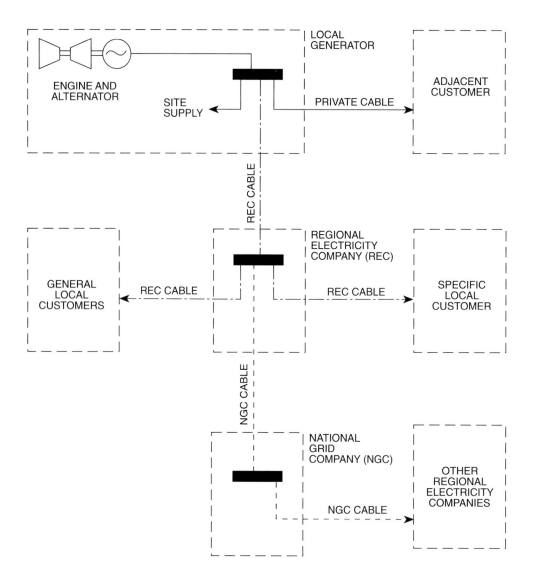

Figure 2.3(a) *Electricity Export Options for a Local Generator*

electricity supplied will be low. Indeed, at certain times of the day prices may be so low as to make the on-site generation of electricity for export not cost effective simply in terms of input fuel costs. The unavoidable conclusion is that at sites where the minimum demands for electricity occur at times of lowest external electricity prices the export of electricity is unlikely to provide a significant contribution to the financing of the CHP plant.

Next, there is the option to sell electricity directly to a National Grid Company in countries where one exists, making use of Regional

Electricity Company cables. Sales of electricity to a National Grid usually entail significant contractual commitments, a high level of uncertainty and the requirement for the exercise of considerable commercial expertise. Despite the higher unit prices that can be secured this is not, therefore, an option that can generally be recommended to the operators of CHP plant. In some countries however, to encourage cogeneration, regulations force the operators of national grids to buy electricity from small generators under favourable contractual arrangements, which include a commitment to purchase output automatically and exclude obligations to generate.

An alternative to the sale of electricity to a Regional Electricity Company or a National Grid Company is to enter into a direct contract to supply a third party, utilising the cables of those companies. In countries where this option is available, the local generator will generally be regarded as an electricity supply company and, therefore, will be required to hold an 'Electricity Supply Licence' like any other electricity utility. The operational demands and legislative requirements associated with a supply licence will usually be quite onerous. In practice, therefore, this option is only likely to be taken up by the operators of large CHP plants, perhaps in excess of 10 MW electrical output (MWe).

Finally, there is the option to supply electricity to an adjacent site. The capital cost of high voltage electrical cabling is low in relation to the value of the energy that can be transported. For this reason, in contrast to the export of heat, the laying of cables to enable the export of electricity to an adjacent site may prove financially attractive, even if that site is many hundreds of metres distant.

In common with the supply of heat, the supply of electricity to a third party may entail significant contractual commitments regarding security of supply. In the case of electricity, however, the customer will almost certainly retain an external supply which can be used in the event of failure of the CHP plant. It may not, therefore, be necessary for the supplier to install standby generating equipment to cover this eventuality. The figures given in table 2.3(a) can be used in conjunction with an estimate of the price currently being paid by an adjacent site for electricity and the distance to that site, to determine the likely financial benefit of exporting electricity directly to a third party.

Like the last option, however, an Electricity Supply Licence will usually be required.

2.3.3 Summary

In conclusion, the export of heat and electricity can provide a valuable source of revenue for a CHP scheme. The option to export heat will, however, be limited to large schemes, perhaps in excess of 10 MWe, where access to low cost capital has been gained and the associated security of supply commitments do not signifi-

cantly add to the burden of operating the CHP plant.

In the case of power, electricity generated in excess of site demand can readily be sold to a Regional Electricity Company with limited capital investment required. In practice, however, as the electricity demand at many sites rises and falls generally in line with total demand on the network of the Electricity Company, excess electricity is most likely to be available for export when the sale price per unit is lowest. The result of this is that, in practice, the generation of electricity by a CHP operator for export to a Regional Electricity Company, will only be worthwhile at times when virtually 100% of the heat also generated can be usefully used.

For similar reasons to heat export, the other options for the sale of electricity can only seriously be considered by operators of large CHP schemes.

2.4 Operation and Maintenance

2.4.1 The Operation and Maintenance Implications

Take an electricity supply from your local electricity company and your operation and maintenance (O&M) commitments to maintain that supply are virtually zero. Equally, use gas in a low temperature hot water boiler to generate heat and maintenance will comprise a simple annual clean and service with the absolute minimum of operational input during the year.

Combined heat and power plant is quite different. Engine and alternator sets will require frequent and costly maintenance work along with regular checking and testing during operation. The commitment of man power and financial resources associated with O&M is significant. Even small scale CHP units are complex in comparison to a boiler and hence place a significant O&M burden upon the operator.

Poor standards of maintenance will be rewarded by poor performance from the CHP plant, with frequent unplanned shut downs and excessive running hours lost for repair work. The result may be that the predicted level of financial benefits for the scheme do not materialise and the project fails to provide an adequate return on investment. In the worst case, the plant may become uneconomic to repair and

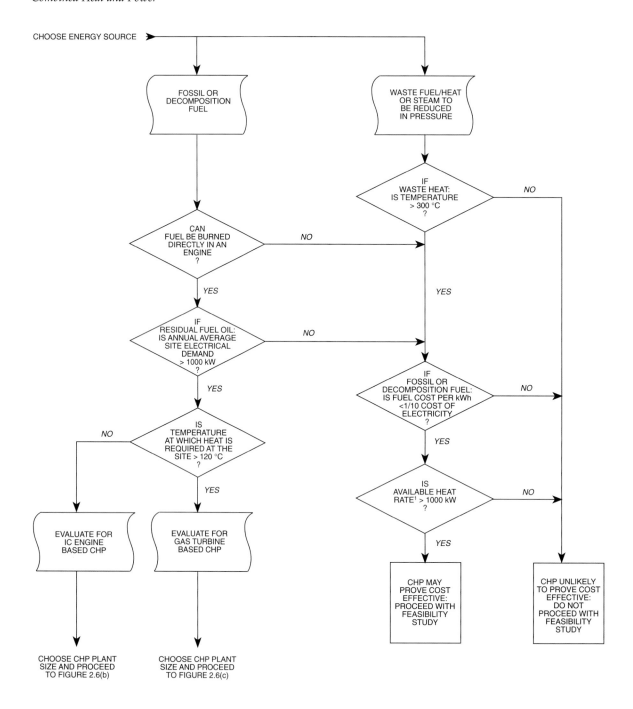

Figure 2.6(a)   *Flow Chart for the Preliminary Evaluation of CHP Scheme Potential*

Notes

1. Available Heat Rate:

Energy Source	Available Heat Rate in kW	Symbols	
Fossil, Decomposition or Waste Fuel	$\dfrac{HHV \times \dot{m}}{6000}$	HHV	– Higher heat value of fuel in kJ/kg
		$\dot{m}$	– Available fuel delivery rate in kg/hr
Waste Heat	$\dfrac{(T-200) \times SH \times \dot{m}}{3600}$	T	– Temperature of waste heat in°C
		SH	– Specific heat of waste heat gases in kJ/kgK
		$\dot{m}$	– Waste heat gas flow rate in kg/hr
Steam	$\dfrac{(h_1 - h_2) \times \dot{m}}{3600}$	h_1	– Enthalpy of supply steam in kJ/kg
		h_2	– Enthalpy of steam at required reduced pressure in kJ/kg
		$\dot{m}$	– Steam flow rate in kg/hr

Figure 2.6(a) continued   *Flow Chart for the Preliminary Evaluation of CHP Scheme Potential*

hence will be taken out of service well short of its predicted operating life.

2.4.2 On-site Staffing Commitments

For most engines, the requirement for regular inspections and checks will necessitate the presence of technicians on-site at frequent intervals. At sites with extensive and complex services, existing plant technicians will be able to undertake the necessary attendance work with a modicum of extra training and with only a small increase in their existing work load. For this type of site the operating costs associated with CHP plant will have only a minimal impact on existing technical staff costs. Many smaller sites, however, have simple services that do not warrant frequent site visits. In these cases, the requirement for a technician to visit site regularly to inspect and check engines, will significantly increase existing technical staff costs.

2.4.3 Maintenance

Engines are complex machines that comprise hundreds of moving mechanical components along with sophisticated electrical and electronic gear. Operating as a base load generating plant, an engine may run in excess of 8,000 hours per annum. For a reciprocating engine, this is equivalent to 400,000 miles in a truck at an average speed of 50 mph, assuming the same engine rpm.

The engines used in CHP plants are either specially designed for static use or have been modified from existing vehicle or aero-engine designs. In either case, when reciprocating engines are used, routine servicing will be required every 1,000 hours or so, i.e. 8 times a year. With modified vehicle engines a major overhaul, including cylinder head exchange, will probably be required once every 18 months. This requirement for frequent maintenance work means that servicing costs for reciprocating engines are high. For a 1 MWe generating set driven by a reciprocating engine, the cost of servicing, repairs and replacement components is likely to be in the region of £70,000 per annum, when the set is run continuously.

As can be seen from table 3.1(b) of chapter 3, this means that approximately 25% of the capital cost of the generating set has to be spent each year on maintenance.

With gas turbines, servicing is based more on a condition monitoring approach and service intervals are longer. Due to the high technology involved, however, actual servicing costs remain substantial at, typically, £45,000 per annum for a 1 MWe gas turbine based generating set.

From the above, it will be realised that the maintenance of CHP plant is a specialised activity requiring expertise in the engine technology concerned. For most organisations, therefore, it is unlikely to be cost effective to utilise in-house staff for this work. Provision will need to be made for the necessary maintenance contracts to be set up with outside contractors.

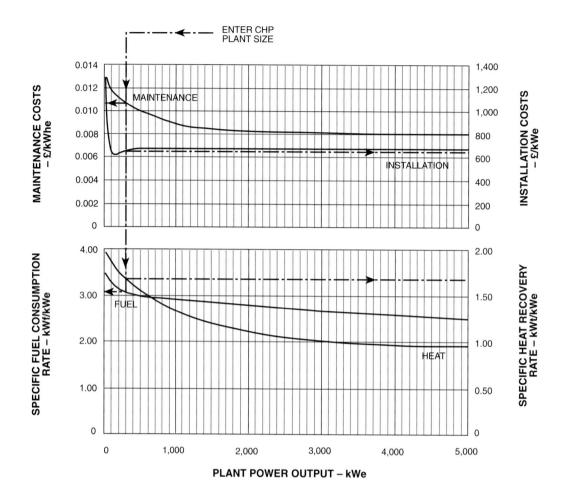

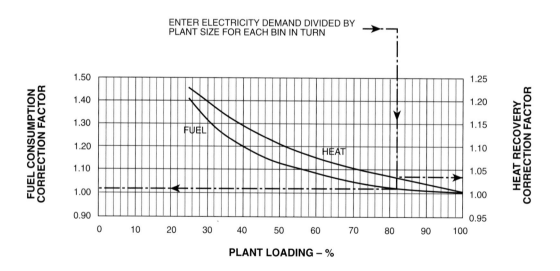

Figure 2.6(b) *Nomograph for the Preliminary Evaluation of CHP Scheme Potential – IC Engines*

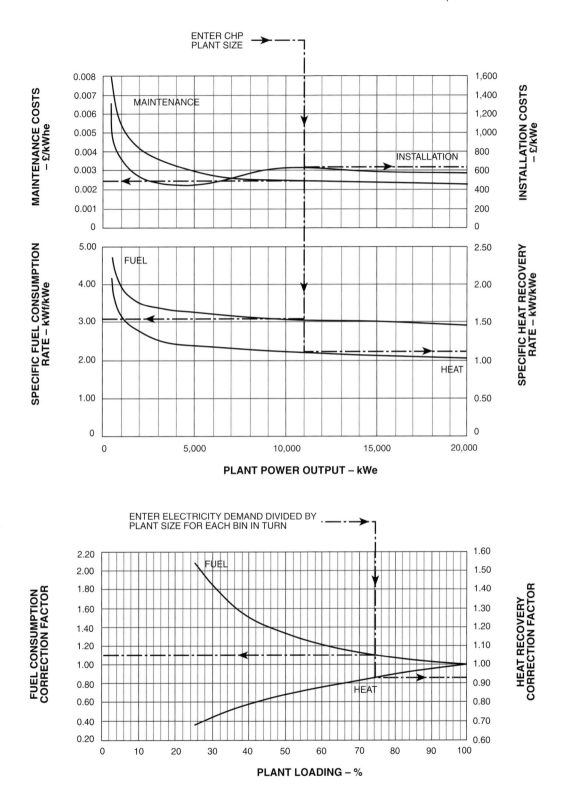

Figure 2.6(c) *Nomograph for the Preliminary Evaluation of CHP Scheme Potential – Gas Turbines*

Table 2.6(a) *Table for the Preliminary Evaluation of CHP Scheme Potential*

Bin	CHP Plant Running Hours[1]	Demand Displaced by CHP Plant – kW		CHP Plant Fuel Demand[4] – kW	Average Energy Unit Prices[5] – £/kWh			CHP Plant Maintenance Costs[6] – £/kWhe	Revenue Savings for the Scheme[7] –£pa
		Electricity[2]	Heat[3]		Electricity	Heat	CHP Fuel		
	A	B	C	D	E	F	G	H	I
1									
2									
3									
4									
Totals									

CHP Plant Installation Costs[8] – £/kWe	CHP Plant Size – kWe	Total Capital Cost for the Scheme[9] –£
J	K	L

Note	Instruction	Item	Source
1	Enter	Number of Weeks × Hours per Week × CHP Plant Availability ÷ 100 (use 90% for reciprocating IC engines and 95% for gas turbines).	Table 2.2(a)
2	Enter the smaller out of	Average Demand for Electricity and CHP Plant Size (kWe)	Table 2.2(a)
3	Enter the smaller out of	Average Demand for Heat and Specific Heat Recovery Rate × Heat Recovery Correction Factor × B (figure from column 'B')	Table 2.2(a) Figure 2.6(b) or 2.6(c)
4	Enter	Specific Fuel Consumption Rate × Fuel Consumption Correction Factor × B	Figure 2.6(b) or 2.6(c)
5	Enter	Weighted average unit prices for fuels, excluding fixed and maximum demand charges. The price for heat should be taken as the price for heating fuel divided by annual overall generating efficiency for the existing boiler plant. In the case of CHP fuel, the price (normally given in terms of the higher heat value of a fuel) will need to be converted to £/kWh at lower heat value.	–
6	Enter	Maintenance Costs	Figure 2.6(b) or 2.6(c)
7	Enter	A× (B×E+C×F – D×G – B×H)	This table
8	Enter	Installation Costs	Figure 2.6(b) or 2.6(c)
9	Enter	J×K	This table

2.4.4 Small Scale CHP and 'All-inclusive Contracts'

At small sites with simple building services, the use of in-house staff to operate and maintain CHP plant is unlikely to be cost effective. At this type of site, the most attractive option for the purchaser will probably comprise an all-inclusive supply and maintenance contract provided by the equipment vendor. The key here is that the purchaser does not increase his existing low O&M commitments but hands over responsibility for the CHP plant to the supplier of the system. The issue of engine O&M demands in relation to site staff is discussed further in chapter 8.

2.4.5 Summary

The operation and maintenance costs associated with combined heat and power plant are high in comparison with simple electrical intake equipment and hot water boilers. It is vital, therefore, that these costs are taken fully into consideration even at the preliminary evaluation stage.

If the existing services at a site are simple, meaning that no technical staff are based at the facility, then the introduction of CHP plant will probably only prove cost effective if both the operation and maintenance of the plant is contracted out.

2.5 Finance

The financing of CHP schemes will be considered in detail in chapter 6. Even at the preliminary evaluation stage, however, it is necessary to consider how the required capital for a CHP project is to be obtained.

CHP schemes rarely provide a simple pay back on investment of less than 3 years. So if internal finance is to be utilised, a pay back criterion in excess of 3 years will almost certainly be required.

CHP equipment suppliers and electricity companies, however, are often interested in funding schemes with pay backs in excess of 3 years as they can obtain capital at relatively low cost and also have long term strategic marketing considerations to take into account. Hence, joint finance or shared savings schemes can sometimes provide the only route to obtaining capital funding for a CHP project.

2.6 Evaluating the Potential for CHP

Now that the engineering and economic environment in which CHP systems have to operate has been sketched out, it is time to consider the potential for cogeneration at a site.

The data compiled for table 2.2(a) should be used in conjunction with the flow chart of figure 2.6(a), the nomographs of figures 2.6(b) and 2.6(c) and table 2.6(a) to execute a preliminary evaluation. A number of CHP plant sizes may need to be tried to achieve the optimum return on capital. A good size to try first, however, is a plant electrical output equal to the overall average site electrical demand for the year.

It will be noted in the calculation of savings that any potential maximum demand charge reductions are ignored. As the purpose of the preliminary evaluation is solely to test the potential for combined heat and power at a site, it is considered that the issue of maximum demand cost savings is a second order consideration, which will not have a fundamental impact on the findings of the preliminary evaluation.

Finally, it must be understood that the costs and savings figures produced from this crude evaluation will only be suitable for use as the basis for a decision on whether to proceed with a feasibility study. *They should never be used as the basis for a capital investment decision.*

3

THE ENGINES
and
ANCILLARY PLANT

At the heart of any combined heat and power plant is the engine. A basic understanding of the characteristics of the various engine technologies available is, therefore, fundamental to the wider understanding of how a CHP plant operates.

THE ENGINES AND ANCILLARY PLANT

3.1 Reciprocating Internal Combustion Engines

For the purposes of this book the term 'internal combustion (IC) engine' will be used to refer to reciprocating IC engines.

3.1.1 Combustion Process
As the term suggests, in an internal combustion engine the burning or combustion of fuel takes place within the engine itself. To be more precise, the fuel is actually combusted in the 'working fluid' used by the engine – in this case air and subsequently air mixed with the products of combustion. This is in contrast to a steam turbine, for example, where fuel is burned in a boiler to raise steam, which is then used as the working fluid of the engine. In the case of the steam turbine, the fuel and products of combustion are kept 'external' to the working fluid by the water tubes of the boiler.

In the pure thermodynamic sense, an internal combustion engine is not a heat engine. Practical designs of IC engines and gas turbines do, however, operate in a manner analogous to heat engines and so, for engineering purposes, can be considered to be heat engines. A more detailed discussion of this fine thermodynamic distinction is given in the first insert panel of this chapter.

Figure 3.1(a) shows the basic components of an IC engine in the form of a cut-away drawing. An illustration of the processes involved in the two types of reciprocating IC engine, 'spark ignition' and 'compression ignition', is given in figure 3.1(b). Finally, photographs of a packaged small scale CHP unit and a 700 kWe CHP installation built around an IC engine are shown in figures 3.1(c) and 3.1(d).

3.1.2 Rejection of Heat
As has been discussed in chapter 1, all heat engines must reject heat in order to generate mechanical work. With an IC engine, heat is

---

1

HEAT ENGINE OR INTERNAL COMBUSTION ENGINE?

As has been discussed in chapter 1, a heat engine takes in heat at a high temperature, converts a proportion of that heat to work and then rejects the remaining energy at a lower temperature. A heat engine must, therefore, by definition operate between a high temperature reservoir and a low temperature reservoir.

An internal combustion engine, on the other hand, takes in energy in the form of the chemical energy held by a fossil fuel, releases that energy through combustion, then converts a proportion of that energy to work whilst rejecting the remaining energy to the atmosphere. An internal combustion engine, therefore, operates with only one, low temperature reservoir – the atmosphere.

In theory, therefore, the thermodynamic laws that apply to heat engines do not apply to internal combustion engines. For example, whilst it is impossible for a heat engine to have an efficiency higher than the Carnot Efficiency, which must always be less than 100%, an internal combustion engine could, in theory, have an efficiency of power generation of 100%.

In reciprocating IC engines and gas turbines, however, the energy released by the combustion of fuel is first transformed into random molecular energy in the working fluid of the engine, thereby raising the temperature of the working fluid. The practical effect, therefore, is equivalent to the transfer of heat to the working fluid from a high temperature reservoir. For this reason, both reciprocating IC engines and gas turbines can be thought of as heat engines, with a theoretical maximum efficiency equal to the Carnot Efficiency.

The efficiency of both reciprocating IC engines and gas turbines is, thus, dependent upon the difference between the maximum temperature achieved in the working fluid through combustion and the temperature at which heat is rejected from the working fluid i.e. the temperature of the exhaust gases. The greater the difference the greater the efficiency.

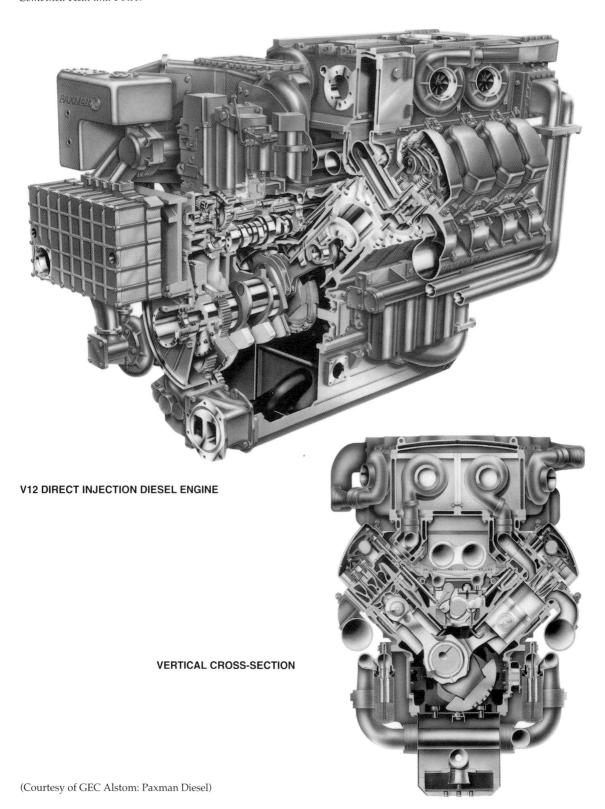

V12 DIRECT INJECTION DIESEL ENGINE

VERTICAL CROSS-SECTION

(Courtesy of GEC Alstom: Paxman Diesel)

Figure 3.1(a) *Cut-away Drawing of a Reciprocating IC Engine*

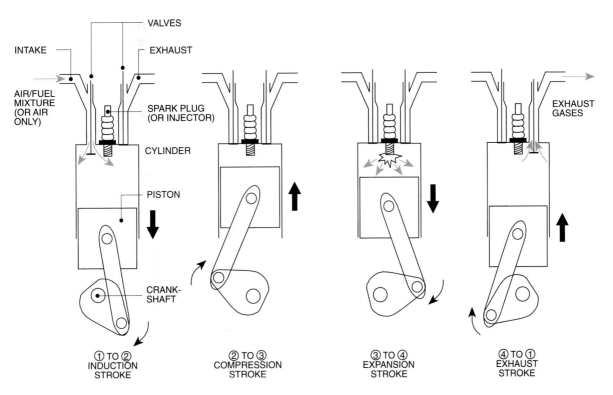

SIMPLIFIED CROSS SECTION THROUGH ONE CYLINDER OF A FOUR STROKE IC ENGINE

Notes

1. Text in (brackets) applies to compression ignition engines.

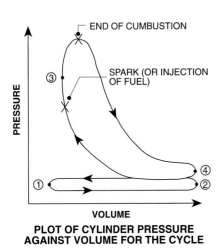

**PLOT OF CYLINDER PRESSURE
AGAINST VOLUME FOR THE CYCLE**

Points in Cycle	Stroke	Operation	
		Spark Ignition	**Compression Ignition**
① to ②	Induction	Intake port open, air/fuel mixture drawn into cylinder.	Intake port open, air drawn into cylinder.
② to ③	Compression	Air/fuel mixture compressed.	Air compressed.
③ to ④	Expansion	Spark initiated ignition and controlled combustion of air/fuel mixture.	Direct injection of fuel into cylinder (or into a small chamber opening onto the cylinder), then spontaneous ignition followed by rapid combustion of air/fuel mixture.
④ to ①	Exhaust	Exhaust port open, products of combustion evacuated from cylinder.	Exhaust port open, products of combustion evacuated from cylinder.

Figure 3.1(b)   *IC Engine Cycles – Spark Ignition and Compression Ignition*

(Courtesy of Nedalo (UK) Ltd)

Figure 3.1(c) *Packaged Small Scale CHP Unit*

(Courtesy of Leverton Power Systems)

Figure 3.1(d) *700 kWe IC Engine CHP Installation*

rejected in the following ways:
a) Directly in the exhaust gases.
b) By heat transfer from the engine jacket water, lubricating oil and charge air (when turbo-charging with aftercooling is employed) cooling systems.
c) By heat transfer from the outer surfaces of the engine.

3.1.3 The Characteristics of IC Engines in Relation to CHP

Power Output For use in CHP plant, IC engines are available from 20 kW to 15 MW power output. From 20 to 100 kW the choice of power plant will generally be limited to vehicle derived engines. For power outputs above 100 kW, engines designed for base load electricity generation are readily available.

Efficiency of Power Generation and Heat Output Efficiency of power generation generally increases as engine size increases and engine speed decreases. Typical figures for engines providing power outputs of 100 kW, 500 kW, 1 MW and 5 MW are given in table 3.1(a). As mentioned above, heat is rejected from IC engines by three mechanisms and is, thus, available at different temperatures. Table 3.1(a) gives typical figures for percentage of input fuel energy and temperature for the heat rejected.

The variation of generating efficiency and

Table 3.1(a) *Typical IC Engine Performance Data*

Item	Percentage of Fuel Input Energy[1]					Temperature – °C	
	Engine Size Range[2] – kW					Flow	Return
	80–120	400–600	400–600 HT[6]	800–1,200	4,000–6,000		
Power Output (variation within each size range)	33 (30–35)	35 (30–38)	35 –	36 (31–41)	41 (35–42)	–	–
Exhaust Gases	22	26	26	26	35	450	–
Engine Jacket Cooling System	33	25	22	24	9	95	85
Lubricating Oil Cooling System	4	4	6	4	4	85	75
Charge Air Cooling System	6	6	6	6	7	35	30
Outer Surfaces of Engine	2	4	5	4	4	–	–
	100	100	100	100	100		
Heat Recoverable at High Temperature[3]	10	11	12	11	16	200	190
Heat Recoverable at Medium Temperature[4]	15	16	39	16	22	120	110
Heat Recoverable at Low Temperature[5]	54	47	47	46	38	80	70

Notes

[1] The figures given in each size range are for a typical spark ignition turbo-charged IC engine operating at full load on natural gas, with input energy taken at the lower heat value.

[2] Refers to shaft power output.

[3] Heat recovered to generate hot water/steam at 200°C, exhaust gas temperature reduced to 230°C.

[4] Heat recovered to generate hot water/steam at 120°C, exhaust gas temperature reduced to 150°C.

[5] Heat recovered to generate hot water at 80°C, exhaust gas temperature reduced to 120°C.

[6] Engine modified to run at high jacket cooling water temperatures of typically 125°C flow and 115°C return.

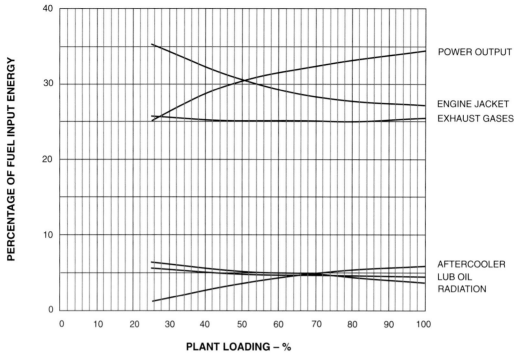

Notes

1. The performance data are for a typical 500 kW shaft power output, spark ignition turbo-charged IC engine operating on natural gas.

Figure 3.1(e) *Typical Part Load Performance – IC Engine*

heat output with engine loading for a typical 500 kW IC engine is given in figure 3.1(e).

Fuels and their Supply A wide range of both liquid and gaseous fuels can be burned in IC engines. In terms of fuel oil, many engines can be operated on light to medium grade oil, with some large marine engines being capable of running on heavy fuel oil. IC engines can also be run on vegetable derived oils. Possible gaseous fuels include natural gas, petroleum derived gases and decomposition gases from sewage works and landfill sites.

Engines run on a clean gas, such as natural gas, will suffer far less from engine deposits and lubricating oil degradation and hence will prove less of a maintenance burden in relation to carburettor, injector and general intake fouling compared to engines run on oil. The associated problem of exhaust valve and valve seat wear is discussed in section 3.1.4.

Where decomposition gases are used, such as from sewage treatment processes, special pre-

cautions and arrangements are required. Firstly, the gases will often contain water in liquid droplet form. This must be removed using a coalescing filter or by drying the gases prior to use in the engine.

Secondly, such gases usually have a lower energy content than natural gas, 22 to 26 MJ/m^3 as opposed to 37 to 39 MJ/m^3 gross calorific value at atmospheric pressure. The higher gas volume flow rates that are, therefore, required to fire the engine necessitate modifications to the fuel metering and delivery systems. Indeed, in some cases, enrichment of decomposition gas using natural gas is necessary to achieve adequate calorific values.

A third issue to consider is the high levels of hydrogen sulphide that occur in gases from sewage treatment works. The presence of hydrogen sulphide in the fuel can cause damage to aftercoolers, engine bearings, bushings, O-rings and gaskets. To prevent such problems occurring, special materials have to be used such as aluminium in the bearings and 'Teflon' in the

O-rings and gaskets. In addition, the products of combustion from engines fired on decomposition gases can be particularly corrosive. Experience has shown that the deleterious effects of such corrosion are minimised when plant is run continuously and where engines are operated with high jacket water temperatures. Crankcase ventilation is also required on engines which burn sewage treatment gases.

Finally, depending upon the exact content of the decomposition gases burned, special materials may have to be used in the construction of silencers, exhaust gas heat recovery equipment and exhaust ductwork.

In liquid form, fuel is relatively easy to introduce into the air intake system or directly into each cylinder using a small pump/diesel injector to provide the necessary pressurisation.

For gaseous fuels, relatively low pressures are required to supply all but the largest of engines. Natural gas at normal low supply pressure, for example, is adequate. For large engines with turbo-charging, where the fuel gas is introduced after the turbo-charger, pressures of up to 3 bar may be required. There is, however, now a trend towards the introduction of gaseous fuel at the inlet to the turbo-charger to obviate the need for high gas supply pressures. In the case of modern dual fuel engines, pressures in the region of 250 bar are used to inject gas directly into the cylinders of these compression ignition type machines. The capital and operating costs associated with the gas compression plant required to deliver the fuel gas supply to dual fuel engines can, thus, be very substantial indeed.

Maintenance and Loss of Service Engine maintenance is considered in detail in chapter 8. Table 8.3(a) of that chapter provides an example maintenance schedule for an IC engine. Naturally, the time required for planned maintenance will reduce the number of hours that the engine is available to run. Unplanned downtime due to faults and component failures will further reduce the available running hours per annum. Typical figures for total engine downtime per annum are given in table 3.1(b).

It should be noted that these figures are for

Table 3.1(b)   *Typical IC Engine based Generating Set Capital and Operating Costs plus Plant Availability*

Item[1]	Generating Set Size[2] – kWe			
	100	500	1,000	5,000
Capital Costs[3] – £	50,000	200,000	300,000	1,800,000
Operating Costs over 10 years[4] :				
Fuel[5] – £	320,000	1,300,000	2,500,000	9,000,000
Maintenance[6] – £	90,000	400,000	700,000	3,200,000
Total Downtime per annum[7] – hours	1,500	1,000	1,000	750

Notes

[1] The base year for all cost figures is 1994.

[2] Refers to generating set electrical power output.

[3] The figures given are for a packaged generating set including: spark ignition turbo-charged IC engine, air cooled alternator, skid, enclosure (except 5,000 kWe set), inlet and exhaust ducts including exhaust silencer, standard control and starting systems plus delivery, testing and commissioning. Gas boosters, flues, heat dump cooling towers, waste heat boilers, electrical switchgear and installation works are excluded.

[4] Operating costs are based on annual running hours of 8,760 minus 'total downtime per annum'.

[5] For the purposes of comparison between engine technologies, nominal fuel prices of 1.2, 1.0, 1.0 and 0.8 p/kWh (at higher heat value) have been assumed for the 100, 500, 1,000 and 5,000 kWe generating sets respectively.

[6] The figures given are based on the cost of a comprehensive warranty and maintenance agreement provided by the manufacturer. This cost includes for all parts, consumables and lubricating oil and all labour such that a purchaser will have to pay no other charges over the 10 year period. The agreement also includes for a breakdown service with response within 24 hours.

[7] Total downtime per annum is the number of hours that the engine is not available to run due to planned maintenance and unplanned outages, based on continuous operation.

well maintained engines designed for continuous operation. Vehicle derived engines are less reliable and require greater maintenance and hence suffer from far greater total downtime per annum. A full discussion of availability and reliability can be found in insert panel 2 of this chapter.

3.1.4 Adaptation of IC Engines for CHP Use

Vehicle versus Static Engine Designs As has been mentioned in chapter 2, for smaller CHP plants use is often made of vehicle derived engines. For a vehicle engine the important performance criteria are:

- High power to weight ratio
- High power to cost ratio
- Reasonable fuel efficiency
- Good reliability

The important criteria for a static IC engine used for base load electricity generation, however, are:

- Reasonable power to cost ratio
- High fuel efficiency
- High reliability
- Long service intervals
- Low maintenance downtime

Unfortunately, the vehicle engine requirement for high power to weight ratio is in conflict with the static engine requirement for long service intervals and low maintenance downtime, as these demand more robust and physically larger engines. Though it is possible to make some modifications to engines designed primarily for vehicle use to make them more suitable for base load generation, such engines are fundamentally not well suited for use in a CHP plant. For this reason, when choosing a power plant for a CHP project, the purchaser should weigh the initial capital cost advantages of a vehicle derived engine very carefully against the significant operating cost advantages of an engine designed for base load generation. The latter will normally provide significant cost savings over the former when life cycle costings are considered.

Ignition In a compression ignition engine, air is compressed in the cylinder, raising its temperature to a level sufficient to cause sponta-

neous combustion when the fuel is injected. Where liquid fuels are used such as oil, spontaneous ignition can be achieved at reasonable temperatures and hence pressures. In the case of a gaseous fuel such as natural gas, however, the extremely high pressure required for spontaneous combustion makes compression ignition impractical. For this reason, when natural gas is used to fuel an IC engine, either spark ignition must be employed or a small quantity of pilot oil must be introduced with the gas to provide compression ignition at reasonable pressures.

IC engines designed to operate on gas with pilot oil are known as 'dual fuel' engines. Dual fuel engines, whilst more expensive to purchase than spark ignition engines, have the advantage that they can be built with the facility to switch over to run on oil alone in the event of a gas supply interruption. A spark ignition engine cannot be operated on fuel oil.

Exhaust Valve Wear Long chain hydrocarbon molecules found in fuel oils tend to form deposits on the back of exhaust valves and on valve seats. These deposits help to prevent high temperature erosion of the metal surfaces as the hot exhaust gases pass from the cylinder to the exhaust system, during the exhaust stroke. When natural gas is used as a fuel, no such deposits occur. The valves and valve seats used in the cylinder heads of gas fired engines have, therefore, to be made of suitably erosion resistant materials. The use of a high ash lubricating oil in gas engines will also provide some coating and hence additional erosion protection of valves and valve seats.

Power Output Control When an engine is used to generate electricity, engine speed has, essentially, to be kept constant to maintain a constant electrical supply frequency. For this reason, speed can not be varied to change power output. Hence, throttling of the air/fuel mixture for spark ignition engines ('quantity governing') and the reduction of oil injected per stroke for compression ignition engines ('quality governing') has to be used to match engine power output to load. With speed constant, frictional losses remain constant whilst power output is reduced. The result is that efficiency of electrical generation falls off as load decreases.

The fall off in efficiency with reducing load is not as marked for compression ignition engines as it is for spark ignition engines. This is due to the fact that the 'quality governing' technique employed on compression ignition engines actually results in improved combustion efficiency at part loads. The throttling of intake air associated with 'quantity governing', on the other hand, lowers volumetric efficiency on the intake stroke which leads to lower combustion efficiency at part load, where spark ignition engines are concerned. Quality governing can not be applied to spark ignition engines, due to the relatively limited range of air/fuel ratios over which ignition with a conventional spark is possible.

From figure 3.1(e) it will be seen that for a typical 500 kW spark ignition engine, efficiency of power generation falls from 35% at full load to 31% and then 25% at one half and one quarter load respectively. The loss of efficiency with reduced load is, thus, marked but does not necessarily preclude the operation of an IC engine at part load in a CHP application, when and if necessary. The significance of this operational flexibility will become apparent when consideration is given to the sizing of engine plant, which is discussed in section 4.5 of chapter 4.

Overall generating set control is discussed in detail in insert panel 8 of this chapter.

Turbo-charging A turbo-charger uses exhaust gases to drive a small turbine which in turn drives a shaft with a compressor at the other end. The compressor is used to compress engine intake air (air/fuel mixture where some spark ignition engines are concerned) thereby getting more working fluid into each cylinder for each cycle. In this way, the power output for a given engine mass can be considerably increased. The normal advantages are: higher power to weight ratio and higher power to cost ratio. The disadvantage is lower low load efficiency.

Turbo-charger performance is improved by cooling the compressed charge air before it passes to the cylinder, using an 'aftercooler'. For effective aftercooling, heat must be removed using cooling water at a relatively low temperature, typically 30 to 50°C. The result is that 10 to 20% of the recoverable heat output from a turbo-charged engine is available at temperatures too low to serve a standard low temperature hot water heating circuit. If use is to be

made of this heat, an alternative application such as ventilation air pre-heating has to be found.

Heat Recovery As has been mentioned in section 3.1.2, heat is available from IC engines from 3 distinct sources: exhaust gases; water, oil and charge air cooling systems and the outer engine casing. To recover heat from the exhaust gases a suitable heat exchanger/waste heat boiler is fitted in the engine exhaust. Due consideration has to be given to the pressure drop across the heat exchanger and its possible effect on engine performance.

The recovery of heat from water or oil cooling systems requires simple water to water and oil to water heat exchangers. The engine jacket water circuit must be kept physically separate from any general purpose heating circuits, as these are unlikely to have the necessary standards of cleanliness and water treatment. As has been mentioned above, in most situations it will not be practical to make use of charge air heat as it is likely to be available at too low a temperature.

Finally, to recover heat from the outer engine casing an enclosure around the engine is required through which air can be drawn, for direct use or to transfer heat to a water circuit via an air to water heat exchanger. As heat can only be recovered by this means at relatively low temperatures, the application of this form of heat recovery is usually limited to ventilation or combustion air pre-heating.

Heat Rejection The engine block and lubricating oil of an IC engine must be kept below certain temperature limits. Hence, whenever the engine is running, heat will need to be rejected from the jacket water and oil cooling systems. Some small CHP installations rely totally on heat recovery to heating circuits for the rejection of this heat. In the event that the demand for heat from the relevant heating circuits falls to less than the required rate of heat rejection from the engine jacket water and oil cooling systems, the engine of such an installation is shut down. Most CHP plants that utilise IC engines are, however, equipped with alternative means of heat rejection in the form of cooling towers. With such provision, the plant can continue to operate to generate electricity when demand for heat at the site is low.

In the case of charge air cooling, the lower the temperature of the cooling water used in the aftercooler, the more effective the charge cooling and the greater the power output from the engine. It is usual to accept that 100% of the heat from charge air cooling will always have to be rejected and, therefore, to select a low aftercooler cooling water temperature to maximise power output from the CHP plant.

Maintenance Intervals and Service Life
Engines that have been designed for continuous operation will already incorporate features, such as large bearing surfaces, to give extended maintenance intervals and a long service life. Such engines require no modifications in terms of maintenance and service life for use in CHP plants.

Engines designed for continuous operation do, however, have a high capital cost. For this reason, many smaller CHP plants make use of low capital cost, vehicle derived power plants. As is mentioned in chapter 2, even with modifications these engines will still require servicing 8 times a year and will require a major overhaul, including cylinder head exchange, every 18 months or so.

In terms of practical service life, vehicle derived engines will often require a complete change of engine block after 20,000 hours of operation. So this type of engine can be considered to have a service life of only 2.5 years when operated continuously. The service life for industrial engines designed for base load generation, however, will usually exceed 100,000 hours or 12 years when operated continuously. Indeed, there are many examples of large IC engines still in service after 200,000 running hours.

The choice between a vehicle derived engine and an engine designed for continuous operation will, thus, be a trade off between capital cost and operating expenditure. It cannot be over emphasised, therefore, that when selecting an IC engine for a CHP application, full life cycle costings must be considered.

3.1.5 Environmental Considerations

Emissions For natural gas firing of IC engines, the exhaust emissions of concern are oxides of nitrogen, carbon monoxide and unburned hydrocarbons. In addition to these emissions, where oil is burned, sulphur dioxide and particulates have to be considered. Finally, the emission of carbon dioxide from engines is an issue in relation to the 'Greenhouse Effect' and global warming. Where hydrocarbon fuels are burned the production of carbon dioxide (CO_2) is inherent to the combustion process. The only practical way to reduce CO_2 emissions, therefore, is to improve the efficiency of engine plant or indeed employ combined heat and power.

Oxides of nitrogen comprise both nitrogen oxide (NO) and nitrogen dioxide (NO_2) and hence are often referred to as 'NO_x'. Oxides of nitrogen are produced at high temperatures during the combustion process, through chemical reaction between oxygen molecules in the combustion air and either nitrogen chemically bound into the fuel or nitrogen molecules also in the combustion air. For this reason, even when fuels which contain no nitrogen are used, such as natural gas, the formation of some NO_x in the cylinder of an IC engine is unavoidable.

The maximum temperature achieved in the cylinder will have a significant impact on the quantity of NO_x formed during combustion. The higher the temperature, the greater the level of NO_x emissions. Unfortunately, engine efficiency improves with increased compression ratios and hence higher combustion temperatures. It should be recognised, therefore, that the achievement of reduced NO_x emissions through the reduction of compression ratio will be at the cost of power generating efficiency (and, incidentally, greater carbon dioxide emissions). In deciding to specify NO_x emission limits below the limits prescribed by local codes or regulations, due note should be taken of this trade-off.

As figure 3.1(f) reveals, where spark ignition engines are concerned, air to fuel ratio also has a major impact on emissions levels as well as determining maximum power output and efficiency. From the plots given, it can be seen that the maximum power from a given spark ignition IC engine is achieved at an air/fuel ratio just below the stoichiometric ratio for the fuel concerned, whereas maximum efficiency is achieved at higher air/fuel ratios. The ratio of actual air to fuel divided by the stoichiometric ratio of air to fuel is sometimes termed 'lambda'.

It will also be clear from the plots that whilst power is maximised at stoichiometric air/fuel

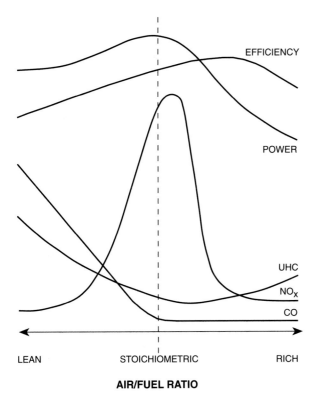

EFFICIENCY

POWER

UHC

NO$_x$

CO

LEAN   STOICHIOMETRIC   RICH

AIR/FUEL RATIO

Notation

CO   – Carbon Monoxide

NO$_x$ – Oxides of Nitrogen

UHC – Unburned Hydrocarbons

Figure 3.1(f) *The Effect of Air/Fuel Ratio on Power, Efficiency and Emissions for a Spark Ignition IC Engine.*

ratios so are NO$_x$ emissions. For this reason, the development of spark ignition engines over the last 10 years has tended to follow one of two paths. 'Lean burn' engines use high air to fuel ratios to minimise NO$_x$, CO and UHC emissions and maximise efficiency at the expense of power output. To regain this loss of power from an engine of a given size, turbo-charging is employed.

The second development path that has been followed is to design spark ignition engines to operate at stoichiometric air/fuel ratios, thereby maximising power output. Exhaust gas treatment equipment is then employed to reduce noxious emissions to acceptable levels before the engine exhaust gases pass to the atmosphere. The form of exhaust gas treatment used on such engines is known as 'non-selective cat-

alytic reduction', as NO$_x$, CO and UHC emissions are all tackled by the treatment.

The process used is based on the ability of rhodium to catalyse the reduction of nitrogen oxide and nitrogen dioxide to oxygen and nitrogen free radicals, at relatively low temperature. The available oxygen free radicals then react with any unburned fuel and carbon monoxide, oxidising them to form carbon dioxide and water. For this 'three-way' catalytic process to work, the right balance of NO$_x$, CO and UHC must be achieved in the exhaust from the engine and no free oxygen must be present at the start of the process. In fact, effective 'three-way catalytic reduction' is typically achieved for lambda values of 0.985 to 0.995. Extremely accurate control of air/fuel ratio is, therefore, required on engines fitted with three-way catalytic converters.

Three-way catalytic converters comprise a ceramic matrix upon which is deposited first a layer of silica and then the catalytic elements, usually rhodium and/or platinum to provide a huge surface area over which the necessary chemical reactions can be catalysed. Over time, however, deposits form on the surface, reducing the area available for catalytic action and hence impairing the performance of the unit. As it is not practical to clean catalytic converters, new units must be fitted to engines on a regular basis. Where an engine is operated for 8,000 hours per annum, catalytic converters may have to be replaced more than once a year.

In situations where lean burn engines are to be used, but NO$_x$ emissions below 50 ppm v/v (at 15% oxygen) are required, then 'selective catalytic reduction (SCR)' of the exhaust gases has to be employed. As the name suggests, SCR tackles only one exhaust gas emission, namely oxides of nitrogen. The technique utilises a reduction agent, usually ammonia (NH$_3$), to react with the oxides of nitrogen to produce N$_2$ and water, in the presence of a catalyst. It should be recognised that ammonia is actually consumed by the process. So, if the carry over of unreduced ammonia in the exhaust gases (known as 'ammonia slip') is to be avoided, close control must be exercised over the injection of ammonia into the catalytic unit.

The requirement for ammonia handling, sophisticated automatic controls and the catalytic unit itself makes selective catalytic reduction a relatively costly technique to employ. As a rough guide, an SCR installation may cost between 50 and 100% of the cost of the engine served and expenditure on ammonia is likely to be in the region of 10% of expenditure on engine fuel. For this reason, SCR is generally only used on engines in the multi-megawatt size range.

A final option available, where lean burn engines are concerned, is the use of 'exhaust gas recirculation (EGR)' to make the exhaust gases from this type of engine suitable for non-selective catalytic reduction. This enables relatively low cost three-way catalytic converters to be used to control exhaust emissions from lean burn engines.

Typical exhaust gas emission levels for a spark ignition engine running on natural gas are given in table 3.1(c).

Turning to compression ignition engines, the high air/fuel ratios used by this type of IC engine lead to the presence of a high level of oxygen in the exhaust gas. Three-way catalytic reduction will not, therefore, work. The options available for the achievement of low NO$_x$ emissions from compression ignition engines include:

- The use of high injection pressures coupled with sophisticated variable ignition timing control to optimise injection duration and timing.
- Water injection.
- Exhaust gas recirculation.
- The treatment of exhaust gases using SCR equipment.

Water injection is a relatively new technique which, whilst effective, is not yet proven in service over a long period of time. Exhaust gas recirculation on compression ignition engines is an effective technique in its own right. Where dirty fuels, such as residual fuel oil, are burned, however, the filtering of exhaust gases before recirculation may be necessary. Where NO$_x$ levels for compression ignition engines are required to be below 70 ppm v/v (at 15% oxygen), the treatment of exhaust gases using selective catalytic reduction equipment will be necessary.

Finally, modern designs of dual fuel engines, sometimes referred to as 'gas diesel engines', have NO$_x$ emissions levels in the region of 230

Table 3.1(c) *Typical IC Engine Exhaust Emissions*

Item	Emission Levels[1]	
	Emission Control Technology	
	Lean Burn	Catalytic Converter[2]
Nitrogen Oxides[3]:		
ppm v/v	50–130	30–130
g/GJ	100–260	60–260
Carbon Monoxide:		
ppm v/v	100–300	70–300
g/GJ	120–370	90–370
Unburned Hydrocarbons[4]:		
ppm v/v	70–180	30–40
g/GJ	50–130	20–30
Particulates:		
mg/nm^3	None	None

Notes

[1] Emission levels are for typical spark ignition engines operating at full load on natural gas. Levels expressed as ppm v/v are for dry gas conditions, adjusted to an exhaust gas oxygen content of 15%.
[2] Standard three-way catalytic converter.
[3] Expressed as NO$_2$
[4] Non-methane.

ppm v/v (at 15% oxygen). With exhaust gas recirculation, however, this type of engine can achieve NO$_x$ emissions below 80 ppm v/v (at 15% oxygen) but with brake thermal efficiencies in excess of 45%. Expressed in terms of work based emissions units (see chapter 5, section 5.9.1), this is equivalent to 1.3 g/kWh which is close to the 1.1 g/kWh level of NO$_x$ emissions achieved by the best lean burn engines.

As compression ignition engines employ high air/fuel ratios, combustion is virtually complete meaning that levels of carbon monoxide and unburned hydrocarbons emissions do not usually present a problem.

Sulphur dioxide and particulate emissions are largely dependent upon the content of the fuel being burned. For this reason, measures to control these emissions centre on the limitation of fuel oil sulphur content and 'aromatic' content, rather than engine or exhaust gas treatment technologies.

Noise Table 3.1(d) gives typical noise levels generated by a 1 MW IC engine. With air filters/silencers fitted, noise emissions from the intakes to an IC engine are relatively low and are not usually a problem. It will be seen from the table, however, that noise emissions from the engine casing, at 90-110 dB(A), are substantial. Exposure to this level of noise can be deleterious to human hearing. For this reason, either ear defenders must be worn by operatives whilst working in an engine room or the engine must be totally enclosed by an acoustic canopy. Acoustic canopies typically provide a sound attenuation of 20 to 25 dB(A).

Unfortunately, IC engines above 1 MW power output are physically large, making acoustic canopies relatively expensive. Larger engines are, therefore, generally not provided with enclosures. Where this is the case, particular attention has to be given to the sound attenuating properties of the surrounding structure, if noise outside the engine room is not to cause a nuisance problem.

The highest levels of noise are emitted from the exhaust of an IC engine. All IC engines are, therefore, fitted with exhaust silencers to attenuate the 100-120 dB(A) of noise that is typically generated at the exhaust manifold.

Vibration As an IC engine involves reciprocating motion by a number of pistons, the transmission of low frequency vibration to surrounding structures can be a problem. Particular attention does, therefore, have to be given to engine mounting arrangements and the connection of all services if the transmission of vibration is to be kept to acceptable levels, where engines are located inside occupied buildings. Under these circumstances, flexible connectors must be used for the connection of all rigid services such as pipes and ducts, and vibration isolation will be required between the steel skid of the engine set and the concrete mounting plinth.

3.1.6 Capital and Operating Costs

Capital Costs The cost of an engine of a certain power output does, of course, vary significantly between modified vehicle engines and engines designed to give low maintenance and a longer service life in a base load generation application. To enable comparisons to be made between engine technologies, however, typical capital costs for packaged generating sets based on IC engines designed for base load operation, have been provided in table 3.1(b). It should be noted that, in common with many items of plant, capital cost versus output reduces markedly as engine size increases. A 500 kW electrical output set, for example, will cost in the region of £200,000 whereas a 1,000 kWe set will cost approximately £300,000. This represents a reduction in cost per unit of power output from £400/kWe to £300/kWe. The relevance of this observation will become clear when the issue of engine plant sizing and the use of multiple engines is considered in chapter 4.

It should be recognised, however, that due to market forces and due to the increased complexity of the ancillary plant associated with larger installations the capital cost of a complete CHP plant may not reflect the engine cost reductions indicated above. This can be seen clearly from the plots of budget CHP installation costs given in figure 2.6(b) of chapter 2.

Operating Costs The costs associated with the operation of IC engine plant comprise the following items:

Fuel
Planned maintenance and repairs
Engine supervision and management

Fuel costs will, of course, be dependent upon the average efficiency of generation achieved by

Table 3.1(d) *Typical IC Engine Noise Emissions*

Source	Noise Emission Levels – dB (A)
Engine Casing[1]	90–110
Intake[2]	100–120
Exhaust[3]	100–120

Notes
[1] Average level measured approximately 1m distant from an engine with no acoustic enclosure.
[2] Measured approximately 1m distant from the turbocharger intake with no air filter/silencer.
[3] Measured approximately 1m distant from the discharge of the engine exhaust manifold.

an engine but will largely be dictated by the type of fuel used and the supply contract negotiated. For the purposes of comparison, however, typical fuel costs have been given in table 3.1(b).

The costs of planned maintenance and repairs are also given in table 3.1(b), based on typical all-inclusive warranty, parts and labour and emergency call out agreements with the engine manufacturer.

Finally, figures for engine supervision and management have not been provided as the wide variation of site circumstances make the compilation of typical figures impossible. In terms of comparisons between engine technolo-

gies the costs will, in any case, be virtually the same.

3.2 Gas Turbines

3.2.1 Combustion Process
In common with IC engines, in a gas turbine fuel combustion takes place within the 'working fluid' used by the engine. This, however, is where the similarity ends. The basic components of a gas turbine are revealed in the cutaway drawing of figure 3.2(a). An illustration of the processes involved is given in figure 3.2(b). Finally, a photograph of a skid mounted gas tur-

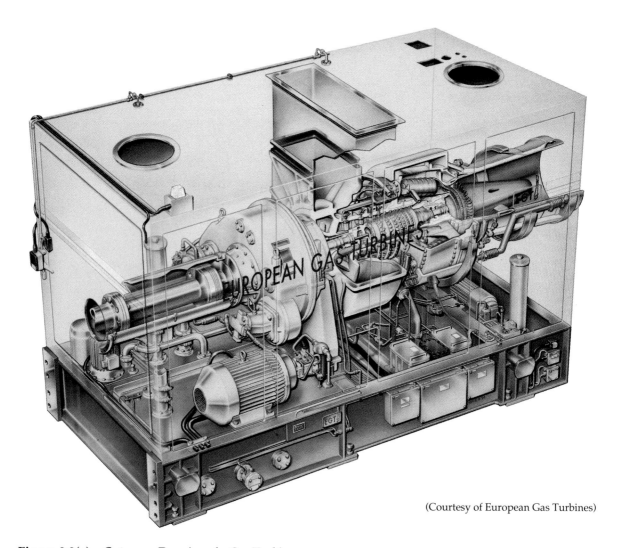

(Courtesy of European Gas Turbines)

Figure 3.2(a) *Cut-away Drawing of a Gas Turbine.*

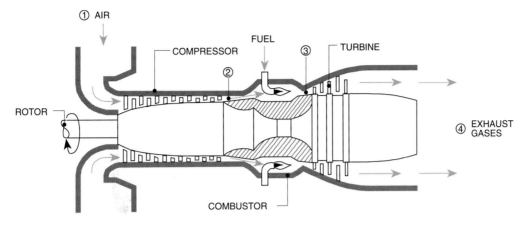

SIMPLIFIED CROSS SECTION THROUGH A TYPICAL SINGLE SHAFT GAS TURBINE

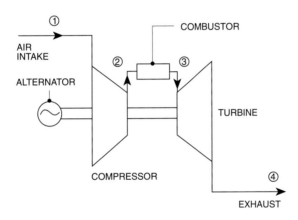

DIAGRAMATIC REPRESENTATION OF
SINGLE SHAFT GAS TURBINE CYCLE

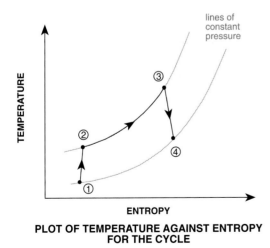

PLOT OF TEMPERATURE AGAINST ENTROPY
FOR THE CYCLE

Points in Cycle	Process
① to ②	Air compressed in axial compressor to typically 14 bar; 400°C.
② to ③	Fuel combusted in compressed air stream at approximately constant pressure to further raise the temperature of the air to typically 1100°C.
③ to ④	Air and combustion products expanded through turbine down to atmospheric pressure and typically 500°C.

Figure 3.2(b)   *Gas Turbine Cycle*

2

RELIABILITY AND AVAILABILITY

Reliability Reliability concerns failure or, to be specific, individual incidents of failure. The reliability of a particular machine or system is, thus, expressed in terms of the expected number of failures over a specified time period. Alternatively, the reciprocal figure is used, 'mean time between failures'. For standby generating sets, reliability is often stated in terms of expected failures to start per 100 starts, i.e. as a percentage.

For CHP installations, however, where return on investment is the principal criterion, reliability figures are of little value as they do not provide information on the likely number of hours lost per annum through engine downtime.

Availability Availability, as the name suggests, concerns the amount of time that a generating set is available to operate (this use of the word should not be confused with 'thermodynamic availability' as discussed in chapter 1). Operating time can be lost for two reasons. Firstly, a generating set may not be available to run due to a failure of the set or the associated installation.

Secondly, running time may be lost when the set is off-line for the execution of planned engine maintenance work. The availability achieved by a generating set in practice will, thus, be dependent not only upon the reliability of the installation but also on the maintenance needs of the engine used.

Availability is often expressed as a percentage from the following formula:

$$A = \frac{H_r - (H_m + H_o)}{H_r} \times 100 \ \cdots\cdots \text{(equation 1)}$$

where
A = Availability
H_r = Number of hours that the plant is required to be in service per annum.
H_m = Number of hours required for planned maintenance per annum (assuming continuous operation).
H_o = Number of hours predicted for unplanned outages per annum (assuming continuous operation).

Manufacturers, however, usually calculate availability in the following way:

$$A = \frac{H_r - (H_m + H_o)}{(H_r - H_m)} \times 100 \ \cdots\cdots \text{(equation 2)}$$

In either case, it will be noted that the availability calculated for a particular generating set changes depending upon the figure chosen for H_r.

Availability Hours For the above reasons, readers are recommended to treat percentage availability figures with extreme caution. To be of any value, a figure for availability must be accompanied by details of the method of calculation. A far more useful concept is the number of hours that a set is available to run per annum. A figure given for 'availability hours' per annum can only be calculated in one way, as follows:

$$H_a = 8760 - (H_m + H_o)$$
where
H_a = Availability hours

Availability Guarantees Manufacturers of engines usually offer 'availability guarantees' as part of a comprehensive warranty and maintenance agreement. The guarantees are usually in the form of an undertaking by the manufacturer to pay the customer an agreed level of damages in the event that the specified availability is not achieved by a generating set. Such agreements often, however, also require the customer to pay to the manufacturer an agreed bonus in the event that the set exceeds the specified availability.

Unfortunately, availability guarantees are usually given in terms of a percentage calculated on the basis of equation 2, given above. This means the impact that the time required for engine planned maintenance has on the availability figure is small. As a reasult, a high percentage availability guarantee can be given by a manufacturer for a poorly designed engine that requires a great deal of maintenance, and hence is unavailable to run for a high number of hours per annum.

bine generating set is shown in figure 3.2(c).

It will be noted that there is a fundamental difference in the way rotary movement is generated by the two types of engine. In an IC engine, combustion takes place in a number of cylinders, where the associated expansion of gases acts on pistons to first generate linear movement. This is then transformed into rotary movement using a number of interconnecting mechanical components all of which have lubricated bearing surfaces which, of course, wear.

In a gas turbine, the products of combustion are expanded through a series of radial blades to generate rotary movement directly. The number of moving mechanical components and bearing surfaces is, thus, significantly less than in an IC engine. It is for this fundamental reason that a gas turbine is more reliable and has far longer service intervals than its IC engine equivalent.

3.2.2 Rejection of Heat

Heat is rejected from a gas turbine in three ways:

a) Directly in the exhaust gases.
b) By heat transfer from the lubricating oil cooling system.
c) By heat transfer from the outer surfaces of the engine.

3.2.3 The Characteristics of Gas Turbines in relation to CHP

Power Output   Gas turbines are available from 200 kW power output to over 200 MW. For CHP

(Courtesy of Centrax Ltd)

Figure 3.2(c) *Gas Turbine Generating Set*

use, however, plants below 1 MW are rarely used due to their poor efficiency and high capital cost per kW of power output. Both aero-derivative and industrial engines are available in the 500 kW to 50 MW range.

Efficiency of Power Generation and Heat Output In common with IC engines, efficiency of power generation increases as the size of gas turbine increases. Typical figures for engines in the 500 kWe to 20 MWe size range are given in table 3.2(a).

Table 3.2(a) also provides typical figures for percentages of input fuel energy and temperature for the heat rejected. From the figures given, it will be noted that the heat is overwhelmingly rejected in the exhaust gases and the temperature at which that heat is available is suitable for raising high pressure steam.

The variation of generating efficiency and

heat output with engine loading for a typical 5 MW gas turbine is given in figure 3.2(d).

Fuels and their Supply Again in common with IC engines, a wide range of liquid and gaseous fuels can be burned in gas turbines. Considering fuel oil first, gas turbines up to approximately 5 MWe require gas oil. Larger sets can, however, operate on medium and in some cases heavy fuel oil. In terms of gaseous fuels, petroleum derived gases, decomposition gases and, of course, natural gas can be utilised.

Maintaining design efficiency in the expansion stage of a gas turbine is, to a significant extent, dependent upon the maintenance of turbine blade cleanliness. For this reason, engines utilising a 'clean fuel' such as natural gas can be operated for many months between turbine washes, with little loss of performance. When fuel oil is used, particularly heavy fuel

Table 3.2(a) *Typical Gas Turbine Performance Data*

Item	Percentage of Fuel Input Energy[1]				Temperature – °C	
	Engine Size Range[2] – kW				Flow	Return
	400–600	800–1,200	4,000–6,000	14,000–24,000		
Power Output	22	27	32	37	–	–
(variation within each size range)	(16–22)	(20–28)	(24–36)	(32–39)		
Exhaust Gases	77	72	67	62	500	–
Lubricating Oil Cooling System	<1	<1	<1	<1	70	50
Outer Surfaces of Engine	<1	<1	<1	<1	–	–
	100	100	100	100		
Heat Recoverable at High Temperature[3]	43	40	38	36	200	190
Heat Recoverable at Medium Temperature[4]	55	52	49	46	120	110
Heat Recoverable at Low Temperature[5]	60	57	53	49	80	70

Notes
[1] The figures given for each size range are for a typical gas turbine operating at full load on natural gas, with input energy based on the lower heat value of the fuel.
[2] Refers to shaft power output.
[3] Heat recovered to generate hot water/steam at 200°C, exhaust gas temperature reduced to 230°C.
[4] Heat recovered to generate hot water/steam at 120°C, exhaust gas temperature reduced to 150°C.
[5] Heat recovered to generate hot water at 80°C, exhaust gas temperature reduced to 120°C.

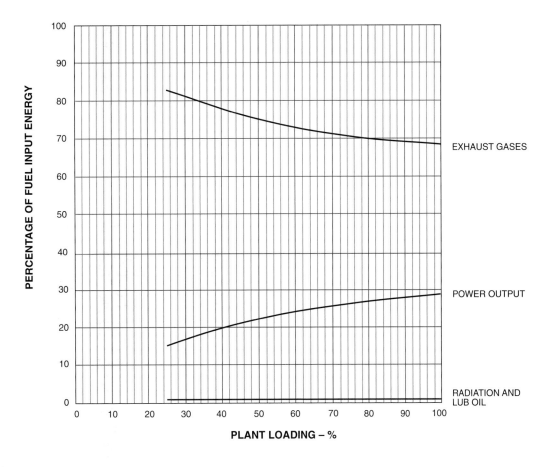

Notes

1. The performance data are for a typical 5 MW shaft power output, gas turbine operating on natural gas.

Figure 3.2(d)   *Typical Part Load Performance – Gas Turbine*

oil, low melting point ashes, unburned hydro-carbons, trace metals and chemicals added to negate the corrosive effect of trace metals may all cause turbine section fouling and lead to the requirement for frequent washes.

Gas turbines require fuel to be supplied to the combustion chamber at high pressure, as the chamber itself is continuously at a high pressure whenever the machine is running. The pressure required is typically from 12 to 20 bar gauge. A rudimentary knowledge of fluid mechanics will remind us that, as liquids are virtually incompressible, little work needs to be done on a fluid in liquid state to pump it to a higher pressure. Hence when a gas turbine is operated only on oil, the cost of the necessary fuel pumping arrangements is low and power losses due to

pumping are negligible. Gaseous fuels, however, are compressible and thus require work to raise them from one pressure to another. In the case of decomposition gases or natural gas available at normal supply pressures, the work required to raise the fuel to a pressure suitable for supply to a gas turbine is significant. For this reason, the cost of the necessary compression plant for a gas turbine operating on natural gas available at district pressures of just 100 mbar, are likely to be in excess of £100,000 for a 5 MWe turbine, with up to 5% of power output being utilised for fuel gas compression.

Now, gas turbine efficiency is related to operating temperature and hence pressure in the power turbine section. The higher the temperature the greater the potential efficiency. But higher power turbine pressures mean higher pressures in the combustion chamber which in turn mean higher fuel supply pressures. When gaseous fuels at low pressure are to be used to fuel a gas turbine, therefore, a trade-off arises between engine efficiency and gas compression losses. For this reason, when use is made of low pressure gas, a CHP plant utilising a high efficiency gas turbine which demands a very high fuel supply pressure may, in practice, be less efficient overall than a plant using a medium efficiency machine that requires more modest fuel pressures. This consideration does not arise when liquid fuels are utilised.

Maintenance and Loss of Service An example maintenance schedule for a gas turbine is given in table 8.3(b) of chapter 8. Typical figures for total engine downtime per annum, both planned and unplanned, are given in table 3.2(b).

A comparison of these figures with those given in table 3.1(b) will reveal that, due to higher reliability and lower maintenance de-

mands, significantly more running hours per annum can be expected from gas turbines than from IC engines. It is, of course, important to consider plant downtime per annum when executing an economic evaluation of a proposed combined heat and power scheme. This is discussed in detail in section 4.7.6 of the next chapter.

3.2.4 Adaptation of Gas Turbines for CHP Use

Aero versus Industrial Gas Turbine Designs
Most gas turbines currently available on the market were originally designed as aircraft propulsion units. For an aero-engine the important performance criteria are:

- High power to weight ratio
- Extremely high reliability
- Good fuel efficiency
- Low maintenance down-time
- Small cross sectional area to minimise drag

Table 3.2(b) *Typical Gas Turbine based Generating Set Capital and Operating Costs plus Availability*

Item[1]	Generating Set Size[2] – kWe			
	500	1,000	5,000	20,000
Capital Costs[3] – £	250,000	400,000	1,200,000	5,000,000
Operating Costs over 10 years[4] :				
Fuel[5] – £	2,200,000	3,600,000	13,000,000	33,000,000
Maintenance[6] – £	350,000	450,000	1,300,000	3,700,000
Total Downtime per annum[7] – hours	500	500	250	250

Notes

[1] The base year for all cost figures is 1994.

[2] Refers to generating set electrical power output.

[3] The figures given are for a packaged generating set including: single fuel gas turbine with no special NO_x control equipment, air cooled alternator, skid, enclosure, inlet and exhaust ducts including exhaust silencer, standard control and starting systems plus delivery, testing and commissioning. Gas compressors, flues, heat recovery boilers, electrical switchgear and installation works are excluded.

[4] Operating costs are based on annual running hours of 8,760 minus 'total downtime per annum'.

[5] For the purposes of comparison between engine technologies, nominal fuel prices of 1.0, 1.0, 0.8 and 0.6p/kWh (at higher heat value) have been assumed for the 500, 1,000, 5,000 and 20,000 kWe generating sets respectively.

[6] The figures given are based on the cost of a comprehensive warranty and maintenance agreement provided by the manufacturer. This cost includes for all parts, consumables and lubricating oil and all labour such that a purchaser will have to pay no other charges over the 10 year period. The agreement also includes for a breakdown service with response within 24 hours.

[7] Total downtime per annum is the number of hours that the gas turbine is not available to run due to planned maintenance and unplanned outages, based on continuous operation.

3

WATER/STEAM INJECTION FOR GAS TURBINES

The technique of injecting water at certain points in the gas turbine cycle is used for two purposes:

- Control of NO_x emissions
- Power and efficiency enhancement

Water/Steam Injection for NO_x Control
Oxides of nitrogen, formed at high temperatures during the combustion of fossil fuels, are generally considered to be a primary contributor to visible pollution and deteriorating air quality. Regulations relating to NO_x emissions from static engines are, thus, virtually universal across the developed world. One method of reducing the formation of NO_x during combustion is to inject small quantities of pure water or steam directly into the combustion chamber. With traditional combustion chamber designs, the injection of water can reduce NO_x emissions by more than 70%, achieving exhaust gas concentrations below 40 ppm v/v (at 15% oxygen).

There are, however, a number of negative aspects associated with water or steam injection. The first concerns emissions of carbon monoxide and unburned hydrocarbons. Emissions of these gases increase as the rate of water/steam injection is increased Secondly, where water is used rather than steam raised by a waste heat boiler located in the gas turbine exhaust, a loss of thermal efficiency will occur. Finally, there is the cost of purchasing and purifying the water used for emissions control. A 5 MW turbine fitted with water/steam injection for emissions control may use up to 5,000m³ of water annually, when running continuously.

Modern low emission designs of combustor can achieve NO_x emissions levels of less than 30 ppm v/v, thereby obviating the need for water/steam injection solely for NO_x control.

Steam Injection for Increased Power Output and Efficiency In addition to NO_x emissions control, steam can be utilised to increase dramatically the power output and efficiency of a gas turbine. Purified water is fed to a waste heat recovery boiler located in the gas turbine exhaust. Using recovered heat from the exhaust gases, steam is raised and then passed to the engine where it can be injected at a number of locations, namely: compressor discharge ports, fuel nozzles (along with the fuel), the combustion chamber and finally at various points along the power turbine casing. The steam expands through the power turbine along with the products of combustion and then passes out of the exhaust to atmosphere.

A gas turbine cycle with steam injection for power enhancement is commonly referred to as the 'Cheng cycle'.

The Cheng cycle improves power output by utilising energy in the exhaust gases, that would otherwise be wasted, to generate more high pressure 'working fluid' to be expanded through the power turbine. The combined mass flow of combustion products and steam can be 14% greater than the combustion products on their own.

However, super-heated steam at the necessary pressure will typically be at a temperature of 500°C, in contrast to the 1100°C temperature of the combustion products in a non-steam injected engine as they pass out of the combustion chamber to meet the blades of the power turbine. The cooling effect provided by the steam enables fuel burn rates to be increased whilst maintaining acceptable temperatures at the power turbine blades.

The combination of increased fuel burn rates and increased mass flow rates enables power outputs for aero-derived gas turbines, in particular, to be increased by up to 60%, with efficiency typically increased by 5% (i.e. 34% to 39%).

Negative aspects of the use of the Cheng cycle include the cost of purchasing and purifying the water that is consumed by the gas turbine (a 5 MW gas turbine with steam injection might consume in excess of 60,000m³ of highly purified water p.a. when running continuously) and patent royalty payments.

The important criteria for land based industrial gas turbines, however, are:

- High reliability
- Good fuel efficiency
- Low maintenance down-time
- Long service intervals

Gas turbines specifically designed for industrial use have been manufactured in small numbers since the 1950's. Many of these units were purchased for oil field and offshore use. For this reason, industrial designs have remained relatively lightweight and compact, keeping the units similar in size to their aero-derived counterparts. Being designed specifically to provide power in rotary form, the compressor, combustion chamber and power turbine configuration have been optimised to provide the best possible performance when driving rotary machinery such as pumps and alternators. In addition, bearing surfaces, casing construction and inspection facilities have all been configured to achieve long service intervals and low downtime. The result is that units can be operated for over 8,000 hours before an extended shut down for detailed inspection and possible maintenance work is required. In the field, on occasions, gas turbines have operated without a break for over 10,000 hours.

Unlike IC engines designed for vehicle use, however, aero-designed gas turbines are essentially already well suited to CHP use. For this reason, the differences between aero-derived gas turbines and static gas turbines should not be considered as fundamental when evaluating alternative engines for a CHP scheme.

Power Output Control As mentioned in section 3.1.4, engines used for electricity generation must operate at a constant speed to maintain electrical supply frequency. In the case of gas turbines, this means that input fuel rate must be varied to match load without changing rotor speed. With speed constant, frictional losses remain constant whilst compression and expansion efficiencies decrease when power output is reduced.

From figure 3.2(d) it will be seen that for a typical 5 MW gas turbine, efficiency of power generation falls from 29% at full load to 23% and then 15% at one half and one quarter load

respectively. It can be seen, therefore, that a gas turbine has significantly less operating flexibility than an IC engine when it comes to part load operation. More stringent constraints are, thus, placed on the sizing of engine plant when gas turbine technology is to be used.

Heat Recovery In common with IC engines, heat is available from three sources: exhaust gases, oil cooling system and the outer engine casing. A gas turbine has no water cooling system. In gas turbine CHP systems, however, heat is only recovered from the exhaust gases, as the quantity and temperature of the heat available from the oil cooling system and engine casing is relatively low.

A waste heat boiler is used to recover heat from the exhaust gases.

Heat Rejection Cooling of power turbine blades is achieved by bleeding off a supply of compressed air from the compressor and passing this via small passageways through the blades. This cooling air passes out of the gas turbine along with the exhaust gas to be lost to atmosphere and hence does not require re-cooling.

Lubricating oil is cooled using a shell and tube oil to water heat exchanger or a direct air blast cooling tower. Finally, heat has to be extracted from the gas turbine acoustic enclosure using a mechanical ventilation system.

Maintenance Intervals and Service Life As mentioned earlier in this section, both aero-derived gas turbine plants and plants originally designed for static use are well suited to cogeneration service, in terms of maintenance intervals and service life.

For gas turbines, maintenance is largely based on a condition monitoring approach rather than on hours run and the results of lubricating oil analysis, as is the case for IC engines. Operating parameters for the machine have to be logged and forwarded to the manufacturer for analysis, on a regular basis. In addition, internal visual inspections of the various 'hot gas' components of the plant (combustion chambers, power turbine blades etc.) are undertaken using a boroscope, thereby obviating the need for regular strip downs. A photograph of a technician undertaking a boroscope inspection of a gas tur-

(Courtesy of Solar Turbines Incorporated)

Figure 3.2(e) *Boroscope Inspection of a Gas Turbine*

bine is shown in figure 3.2(e).

Every 8,000 hours, however, some components may require stripping down for closer visual inspection. After typically 40,000 hours an overhaul takes place which may include the fitting of exchange hot gas and rotating components.

As far as overall service life is concerned, it is not unusual to find gas turbine plants still in operation after 100,000 hours of use i.e. 12 years of continuous operation. Some plants have recorded in excess of 200,000 operating hours.

3.2.5 Environmental Considerations

Emissions Figures for typical exhaust emissions are given in table 3.2(c). The reduction of exhaust emissions through the use of water/steam injection has been discussed in insert panel 3 earlier in this chapter. The latest combustor designs are, however, equally if not more effective than water injection alone in controlling NO_x emissions. These designs achieve

Table 3.2(c) *Typical Gas Turbine Exhaust Emissions*

Item	Emission Levels[1]		
	Emission Control Technology		
	None	Water Injection	Low NO_x Burners
Nitrogen Oxides[2]:			
ppm v/v	80–170	20–50	20–30
g/GJ	160–340	40–100	40–60
Carbon Monoxide:			
ppm v/v	10–30	10–60	10–50
g/GJ	10–40	10–70	10–60
Unburned Hydrocarbons[3]:			
ppm v/v	0–10	10–20	10–20
g/GJ	0–10	10–15	10–15
Particulates:			
mg/nm³	None	None	None

Notes
[1] Emission levels are for typical gas turbines operating at full load on natural gas. Levels expressed as ppm v/v are for dry gas conditions, adjusted to an exhaust gas oxygen content of 15%.
[2] Expressed as NO_2.
[3] Non-methane.

lower emissions by reducing the localised areas of high temperature or 'hot spots', where the bulk of NO_x is formed, by improving pre-mixing of fuel and air and through better flame control.

In situations where particularly stringent regulations apply for NO_x and other emissions (i.e. NO_x less than 20 ppm v/v at 15% oxygen), exhaust gas treatment systems may be required. For carbon monoxide control a carbon monoxide reducing catalyst or catalytic converter is required. In the case of oxides of nitrogen, 'selective catalytic reduction' units in which ammonia is injected into the exhaust gas stream in the presence of a catalyst, are used to break NO_x back into free nitrogen and oxygen. The technique of selective catalytic reduction has been discussed at some length in section 3.1.5. When applied to gas turbines the technique can reduce NO_x emissions levels to 10 ppm v/v (at 15% oxygen).

Table 3.2(d)   *Typical Gas Turbine Noise Emissions*

Source	Noise Emission Levels – dB (A)
Engine Casing[1]	100–110
Intake[2]	130–140
Exhaust[3]	120–130

Notes

[1] Average level measured approximately 1m distant from a gas turbine with no acoustic enclosure.

[2] Measured approximately 1m distant from the compressor intake with no air filter/silencer.

[3] Measured approximately 1m distant from the exhaust centre-line, on the plane of the outlet flange.

Noise   The levels of noise generated by a typical 5 MW gas turbine are given in table 3.2(d). In contrast to an IC engine, noise emitted from the intake to a gas turbine is potentially a problem, as the intake air is ducted directly from outside. The requirement to minimise intake system pressure drop, however, means that particular care is taken in the design of intake noise attenuation for gas turbines.

As gas turbines are physically compact in relation to their power output, all but the very largest of sizes are supplied with acoustic enclosures as a matter of course.

Vibration   Unlike reciprocating IC engines, only rotary movement is involved in a gas turbine. For this reason, the transmission of vibration from gas turbines to surrounding structures is not a problem.

3.2.6 Capital and Operating Costs

Capital Costs   Even with the globalisation of the gas turbine market, the capital cost of machines with similar power outputs can vary markedly from manufacturer to manufacturer. To enable comparisons between engine technologies, however, typical capital costs for packaged generating sets based on gas turbines have been provided in table 3.2(b). A comparison with the figures given in table 3.1(b) will show that a 1 MWe gas turbine based set could be expected to cost around 30% more than an equivalent IC engine based set, in terms of ex-works prices.

Operating Costs   The costs associated with the operation of gas turbine plant comprise:

Fuel
Planned maintenance and repairs
Engine supervision and management

Fuel costs over 10 years for the same nominal gas prices used for table 3.1(b), are given in table 3.2(b). The consequence of the significantly lower fuel to power efficiency of gas turbines in comparison with IC engines can clearly be seen from the figures in the tables.

Finally, typical costs for planned maintenance and repairs are also given in table 3.2(b). The advantages provided by gas turbines over IC engines can be seen from a comparison of these figures with those given in table 3.1(b).

3.3 Steam Turbines

3.3.1 Combustion Process

Unlike reciprocating IC engines and gas turbines, in the case of steam turbines, fuel combustion takes place outside the working fluid. The necessary energy is imparted to the working fluid, in this instance steam, by boiling off water vapour in a heat exchanger which keeps combustion products and water separate i.e. a steam boiler. Figure 3.3(a) reveals the basic components of a steam turbine. An illustration of the processes involved is given in figure 3.3(b). Finally, a photograph of a steam turbine based generating set is shown in figure 3.3(c).

It will be informative to contrast the operation of a steam turbine with that of a gas turbine. In a gas turbine the working fluid, air, is first compressed to roughly 14 bar pressure using significant shaft power to drive a radial compressor. Fuel is then combusted within the air to raise the temperature of the working fluid from perhaps 400°C to 1,100°C, at approximately constant pressure. This high temperature, high pressure fluid is then expanded through a series of radial turbine blades to generate shaft power. Any shaft power generated that is in excess of the power required to drive the compressor, is available to drive external machinery.

For a steam turbine, however, the working fluid starts off as a liquid in the form of water. A feed water pump is used to pump the liquid at high pressure into a steam boiler. As the working fluid is, at this stage, a virtually incompressible liquid, very little energy is required to transfer the fluid from low to high pressure. This is in complete contrast to the compression of air in the gas turbine.

In the boiler of the steam turbine plant, however, comparatively greater energy must be transferred to the working fluid as the fluid has to be first changed from a liquid to a gas then the pressure and temperature of the gas must be increased, all through the addition of heat. The temperature and pressure of the steam raised would, typically, be in the region of 200-500°C and 10-60 bar respectively. The steam is then expanded through a series of fixed and rotating radial turbine blades to generate shaft power, all of which is available to drive external machinery.

In both gas turbines and reciprocating IC engines the working fluid is exhausted to atmosphere after it has been utilised in the engine one time. The water used in a steam turbine, however, is recovered from the turbine exhaust, condensed and then returned in liquid form to the feed water pump to be used again and again. A steam turbine system can, thus, be thought of as operating on the basis of a complete thermodynamic cycle with heat transfer to the working fluid at high and low temperatures and may, therefore, be considered to be a true heat engine.

In common with gas turbines, rotary movement in steam turbines is generated directly, keeping moving mechanical components and bearing surfaces to a minimum. Steam turbines are, thus, highly reliable and have long service intervals.

3.3.2 Rejection of Heat

Heat is rejected from a steam turbine through three mechanisms:

a) By heat transfer to site equipment or to a condenser, to turn the steam to liquid water before returning it to the feed water pump.
b) By heat transfer from the lubricating oil cooling system.
c) By heat transfer from the outer surfaces of the engine.

3.3.3 The Characteristics of Steam Turbines in relation to CHP

Power Output Steam turbines are available from less than 100 kW power output to over 500 MW.

4

STEAM RATE AND EFFICIENCY FOR STEAM TURBINES

Steam Rate As steam turbines do not convert fuel to power directly, the term 'brake thermal efficiency' is not relevant to this type of engine. It is usual practice to define the performance of steam turbines in terms of 'steam rate', which is the quantity of steam needed to produce a certain quantity of shaft work. The units commonly used are kg/kWh.

For steam supplied to a steam turbine at condition 1 and exhausted at condition 2, the 'theoretical steam rate' is given by:

$$w = \frac{3600}{h_1 - h_2}$$

where
w = theoretical steam rate (kg/kWh)
h_1 = enthalpy (kJ/kgK) of steam at condition 1 pressure and temperature
h_2 = enthalpy (kJ/kgK) of steam at condition 2 pressure but condition 1 entropy

Steam Turbine Efficiency The actual steam rate for a steam turbine is greater than the theoretical steam rate due to thermodynamic losses in the conversion of steam pressure to rotary movement and due to mechanical losses. The 'efficiency' of a steam turbine is defined as:

$$Efficiency = \frac{Theoretical\ Steam\ Rate}{Actual\ Steam\ Rate} \times 100$$

Typical efficiencies for steam turbines range from 55% to 80%.

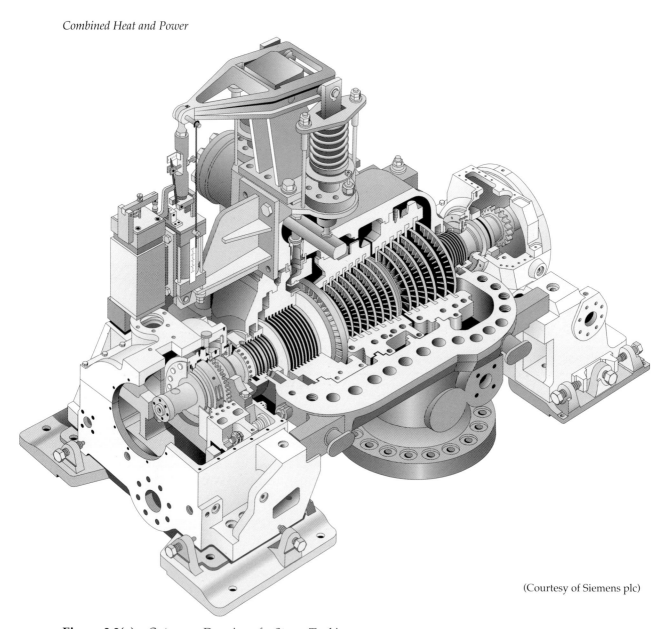

(Courtesy of Siemens plc)

Figure 3.3(a)   *Cut-away Drawing of a Steam Turbine*

Efficiency of Power Generation and Heat Output Efficiency increases with increased turbine blade speed. Hence, large diameter turbines operating at high rotational speeds give the highest efficiencies. For a single machine, however, the efficiency against load characteristic is not fixed but varies with steam inlet and outlet conditions. The greater the difference between inlet and outlet pressure and the higher the degree of supply steam superheat, the greater the power generating efficiency of the turbine. Where steam exhausted from the turbine is to be used for a second function, such as heating, outlet pressure will need to be above atmospheric. For such an application a 'back-pressure' design of steam turbine will be used. In cases where the maximum work is to be extracted from a given steam supply, outlet pressure will be held as far below atmospheric pressure as practicably possible and a 'condensing' design of steam turbine will be used.

Typical efficiency figures for a nominal set of

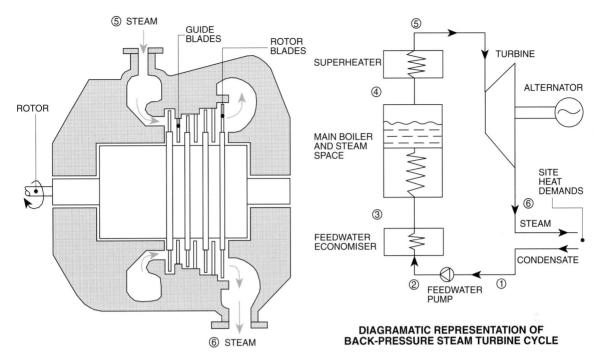

SIMPLIFIED CROSS SECTION THROUGH A TYPICAL
BACK-PRESSURE STEAM TURBINE

DIAGRAMATIC REPRESENTATION OF
BACK-PRESSURE STEAM TURBINE CYCLE

Points in Cycle	Process
① to ②	Condensate pumped from low pressure to high pressure to become boiler feedwater.
② to ③	Feedwater passed through economiser to raise temperature.
③ to ④	Water heated in boiler tubes, passed to steam space and dry saturated steam separated off.
④ to ⑤	Saturated steam passed through superheater to generate superheated steam.
⑤ to ⑥	Superheated steam expanded through steam turbine to superheated or saturated steam at chosen back-pressure.
⑥ to ①	Steam at chosen back-pressure used for heating at site, condensed and reduced to atmospheric pressure, then pumped back to boiler house as condensate.

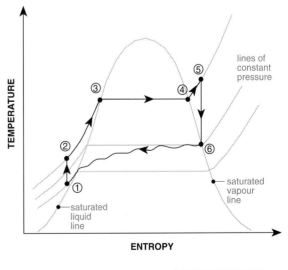

PLOT OF TEMPERATURE AGAINST ENTROPY
FOR THE CYCLE

Figure 3.3(b)   *Steam Turbine Cycle*

(Courtesy of Peter Brotherhood Ltd)

Figure 3.3(c) *Steam Turbine Generating Set*

supply and exhaust conditions are given in table 3.3(a) for back-pressure turbines. Condensing turbines are covered by table 3.3(b). These two tables also give heat rejection figures for the two types of machine operating at the chosen, nominal inlet and outlet conditions. The variation of generating efficiency and heat output with engine loading for a typical 1 MW back-pressure steam turbine, is given in figure 3.3(d).

Fuels and their Supply As the steam turbine cycle does not involve 'internal combustion', the products of combustion do not have to pass through the engine. Provided a suitable means of fuel feed, fuel burn and waste product rejection can be devised, virtually any combustible material can be used to fire a steam turbine plant. As reciprocating IC engines and gas tur-

bines have higher power generating efficiencies, however, it does not make economic sense to utilise fuels that could be directly used in these types of machine to fire the boiler of a steam turbine plant.

For this reason, where a steam turbine is to be employed as the main heat engine of a CHP system either coal, heavy fuel oil or combustible wasted material, such as refuse, is usually burned to raise steam. Even in the case of heavy oil, however, large IC engines and gas turbines are manufactured that can operate directly on the fuel.

The 'fuel in' to 'power out' efficiency of the overall plant is, of course, dependent not only on the efficiency of the turbine but also on the steam generating efficiency of the boiler. As low cost and hence 'dirty' fuels are burned in the boilers of steam turbine CHP systems, regular cleaning of boiler flue-ways is of particular importance to the maintenance of acceptable combustion efficiencies for this type of plant.

Maintenance and Loss of Service   An example maintenance schedule for a steam turbine is given in table 8.3(c) of chapter 8. Typical figures for total engine downtime per annum, both planned and unplanned, are given in table 3.3(c). A comparison of these figures with those for IC engines and gas turbines will show that steam turbines are the most reliable and least maintenance intensive prime movers available.

3.3.4 Adaptation of Steam Turbines for CHP Use

Due to the extremely high pressures that can be involved, the casings of steam turbines tend to be substantial and therefore heavy. In addition, as combustion takes place outside the working fluid, a large surface area is required to transfer heat from the combustion products to the steam. Hence, a relatively bulky boiler is required. For these reasons, steam turbines are exclusively designed for static use or marine applications and thus require little modification for application to CHP.

Exhaust Steam Pressure   There is, however, one important item that requires detailed consideration and that is exhaust steam pressure.

From the discussion of heat engines given in section 1.2, it will be remembered that the higher the temperature at which heat is rejected from an engine the lower the efficiency of power generation. In the case of gas turbines and IC engines, the various temperatures at which heat

Table 3.3(a)   *Typical Back-pressure Steam Turbine Performance Data*

Item	Percentage of Fuel Input Energy[1]				Temperature – °C	
	Engine Size Range[2] – kW				Flow	Return
	400–600	800–1,200	4,000–6,000	14,000–24,000		
Power Output	5	5	6	7	–	–
Exhaust Steam	68	68	68	67	160	–
Lubricating Oil Cooling System	<1	<1	<1	<1	70	50
Outer Surfaces of Engine	<1	<1	<1	<1	–	–
Steam Generating Losses	25	25	25	25	–	–
	100	*100*	*100*	*100*		
Heat Recoverable at High Temperature[3]	0	0	0	0	200	190
Heat Recoverable at Medium Temperature[4]	66	66	65	64	120	110
Heat Recoverable at Low Temperature[5]	68	68	68	67	80	70

Notes

[1] The figures given in size ranges 400–600 and 800–1,200 are for typical single stage back-pressure steam turbines, whereas those given in the 4,000–6,000 and 14,000–24,000 size ranges are for multi-stage back-pressure machines. In each case, the turbines are operating on superheated steam at 10 bar (absolute), 300°C and an exhaust pressure of 3 bar (absolute). It has been assumed that steam is raised in a conventional gas fired boiler plant operating at an annual average overall generating efficiency of 75%.

[2] Refers to shaft power output.

[3] Steam is exhausted at a temperature of approximately 160°C, which is too low to generate hot water at 200°C.

[4] Steam used directly or to generate hot water at 120°C.

[5] Steam used to generate hot water at 80°C.

Table 3.3(b) *Typical Condensing Steam Turbine Performance Data*

Item	Percentage of Fuel Input Energy[1]				Temperature – °C	
	Engine Size Range[2] – kW				Flow	Return
	400–600	800–1,200	4,000–6,000	14,000–24,000		
Power Output	11	11	15	16	–	–
Exhaust Steam	63	63	59	58	70	–
Lubricating Oil Cooling System	<1	<1	<1	<1	70	50
Outer Surfaces of Engine	<1	<1	<1	<1	–	–
Steam Generating Losses	25	25	25	25	–	–
	100	100	100	100		
Heat Recoverable at High Temperature[3]	0	0	0	0	200	190
Heat Recoverable at Medium Temperature[3]	0	0	0	0	120	110
Heat Recoverable at Low Temperature[3]	0	0	0	0	80	70

Notes

[1] The figures given in size ranges 400–600 and 800–1,200 are for typical single stage condensing steam turbines, whereas those given in the 4,000–6,000 and 14,000–24,000 size ranges are for multi-stage condensing machines. In each case, the turbines are operating on superheated steam at 10 bar (absolute), 300°C and an exhaust pressure of 0.3 bar (absolute). It has been assumed that steam is raised in a conventional gas fired boiler plant operating at an annual average overall generating efficiency of 75%.

[2] Refers to shaft power output.

[3] Exhaust pressure (and hence temperature) used for condensing steam turbines is maintained at a low level to maximise power output and efficiency. At the figure used for this table, of 0.3 bar (absolute), saturated steam has a temperature of approximately 70°C. Hence, it is not possible to recover heat to generate hot water at either 200, 120 or 80°C. In fact, cooling towers operating with an off-tower water temperature of 60°C are required.

is rejected from the engines are pretty much fixed at the time of design. The CHP system designer thus has to make do with the available heat at the temperatures given. This is not so in the case of steam turbines. Back-pressure steam turbines are designed to be supplied with steam at a range of pressures and to exhaust steam at a range of pressures above atmospheric pressure. The CHP system designer does, therefore, have some flexibility in the choice of supply and exhaust pressures for steam turbines. A trade off has to be made between high supply/low exhaust steam pressures to give optimum power efficiency and low supply/high exhaust steam pressure to keep boiler, pipework, heat exchanger and steam turbine costs down.

At a site that has a requirement for steam at 2 or more different pressures, steam can be taken directly from the boiler to serve the high pressure application and from the turbine exhaust to serve the low pressure application. An alternative is to have a steam turbine with a number of turbine blade sections and to exhaust a proportion of the steam passing through the engine at some point in between the turbine sections. This type of steam turbine is known as a 'back-pressure pass-out' turbine.

Finally, in situations where heat is required below 100°C or is not to be recovered, steam can be exhausted from the turbine below atmospheric pressure leading to immediate condensation into liquid form. In this application a condensing steam turbine is used.

Table 3.3(c) *Typical Steam Turbine based Generating Set Capital and Operating Costs plus Availability*

Item[1]	Generating Set Size[2] – kWe			
	500	*1,000*	*5,000*	*20,000*
Capital Costs[3] – £	100,000	200,000	1,500,000	3,000,000
Operating Costs over 10 years[4] : Fuel[5] – £	9,000,000	18,000,000	61,000,000	208,000,000
Maintenance[6] – £	30,000	50,000	400,000	800,000
Total Downtime per annum[7] – hours	200	200	100	100

Notes

[1] The base year for all cost figures is 1994.

[2] Refers to generating set electrical power output.

[3] The figures given are for a packaged generating set including: back-pressure steam turbine (single stage for 500 and 1,000 kWe sets, multi-stage for 5,000 and 20,000 kWe sets), air cooled alternator, skid, enclosure, standard control and starting systems plus delivery, testing and commissioning. Boiler plant, electrical switchgear and installation works are excluded.

[4] Operating costs are based on annual running hours of 8,760 minus 'total downtime per annum'.

[5] For the purposes of comparison between engine technologies, nominal fuel prices of 1.0, 1.0, 0.8 and 0.6p/kWh (at higher heat value) have been assumed for the 500, 1,000, 5,000 and 20,000 kWe generating sets respectively.

[6] The figures given are based on the cost of a comprehensive warranty and maintenance agreement provided by the manufacturer. This cost includes for all parts, consumables and lubricating oil and all labour such that a purchaser will have to pay no other charges over the 10 year period. The agreement also includes for a breakdown service with response within 24 hours.

[7] Total downtime per annum is the number of hours that the steam turbine is not available to run due to planned maintenance and unplanned outages, based on continuous operation.

Direct Steam Turbine Drives The option to use a steam turbine directly to drive an item of plant such as a centrifugal chiller rather than to generate electricity, should not be overlooked. Such an application is relatively unusual in the UK but not uncommon on larger plant in the US. One advantage is that the losses associated with the generation of electricity and its conversion back into rotary movement are eliminated. Perhaps more important, however, is the flexibility to provide rotational speed control that becomes possible through the use of a directly driven plant.

Part load efficiency for steam turbines is significantly improved when speed can be reduced at lower loads. In addition, the part load efficiencies of centrifugal refrigeration and pumping equipment is improved when variable speed control is utilised. Where a steam turbine is used to drive pumping or refrigeration equipment directly, therefore, part load efficiency for the overall plant will be superior to that for an equivalent plant operating at constant speed to generate electricity, which then powers electric motor drives on the equipment served.

Power Output Control As has already been mentioned, engines used for electricity generation must operate at a constant speed. To maintain constant speed at varying loads, the steam supply to a steam turbine is throttled using a valve or series of valves in the supply. In common with gas turbines, with speed constant, frictional losses remain constant whilst expansion efficiency decreases when power output is reduced. Depending upon the design of turbine and the inlet and outlet steam conditions under consideration, the reduction in efficiency with falling load may be modest (as shown in figure 3.3(d)) or may be significant.

When used as a direct drive for centrifugal chillers or pumps, speed control is not only possible but is desirable to optimise the part load efficiency of the equipment served. Under these circumstances, speed is set by the automatic controls of the equipment served in response to load on that equipment. The steam supply to the steam turbine is then throttled to maintain the required rotational speed. With speed falling as load reduces, frictional losses decrease whilst expansion efficiency is more nearly maintained

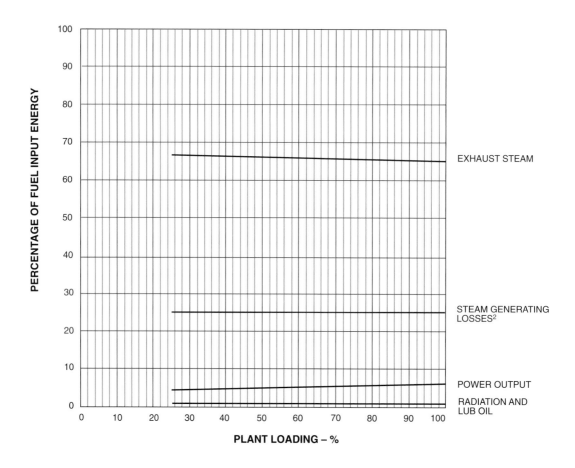

Notes
1. The performance data are for a typical 1 MW shaft power output, back-pressure steam turbine.
2. A nominal figure of 25% has been assumed for boiler plant steam generating losses.

Figure 3.3(d) *Typical Part Load Performance – Steam Turbine*

at its optimal value. It is for these reasons that the part load performance of steam turbines can be improved when speed control is admissible.

Heat Recovery As is stated in section 3.3.2, heat is potentially available from three sources: exhaust steam, oil cooling and the outer engine casing. Like a gas turbine, a steam turbine has no water cooling system and heat recovery from the oil cooling system and from the engine casing is usually not undertaken as the quantities of heat available are relatively small.

Unlike the exhaust from a gas turbine, however, the exhaust from a steam turbine is already in the form of steam ready to be used directly for process or space heating. A further heat exchanger in the exhaust is not, therefore, required. The implications of the pressure chosen for the steam turbine exhaust have been discussed earlier in this section.

Heat Rejection Lubricating oil is cooled using a shell and tube oil to water heat exchanger or a direct air blast cooling tower. In the case of condensing turbines, the exhaust steam is passed to a shell and tube heat exchanger to be cooled by a water circuit, which in turn rejects heat at cooling towers.

5

COMBINED CYCLE PLANTS

In the conventional application of steam turbines to CHP the turbine is used as the main heat engine of the plant. An alternative and important use for steam turbines, however, has been rapidly developing in recent years. This second application concerns the squaring up of the heat to power output ratio for a CHP system.

At sites which have a heat to power ratio of less than 3:1, there may be insufficient demand for the heat generated by a straightforward gas turbine installation. Under these circumstances, it can sometimes be cost effective to utilise a proportion of the heat rejected by the gas turbine to raise steam at sufficient pressure to drive a steam turbine, thereby producing more power and less heat from the system. In this way, a better match can be achieved between site demands and plant output.

As this type of system configuration combines the open cycle gas turbine process with the closed cycle of a steam turbine, it has become known as 'combined cycle plant'. A typical combined cycle plant configuration is shown diagrammatically in figure 3.3(e).

The primary function of the steam turbine in this application is to generate further power. Exhaust pressures will, therefore, be kept as low as possible to maximise power generating efficiency. For this reason, condensing steam turbines are used in combined cycle plants.

In the case of gas turbine power stations, where no heat is to be recovered, the use of steam turbines can raise overall electrical generating efficiency from under 35% for a gas turbine only plant to over 50% for a combined cycle plant. A drawing showing the main components of a typical 'combined cycle gas turbine (CCGT)' power station is shown in figure 3.3(f).

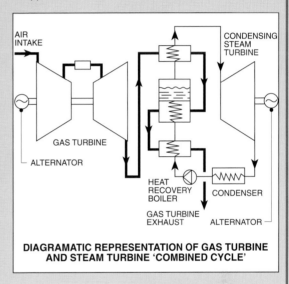

DIAGRAMATIC REPRESENTATION OF GAS TURBINE AND STEAM TURBINE 'COMBINED CYCLE'

Figure 3.3(e)   *Combined Cycle*

Maintenance Intervals and Service Life
Steam turbines are relatively simple engines with a limited number of moving parts. With the exception of lubricating oil and oil filter changes, maintenance intervals can be as long as 25,000 hours. In common with gas turbines, maintenance is essentially based on a condition monitoring approach, with components being replaced when clearances etc. are no longer within limits.

As regards overall service life, this may be considered as virtually indefinite, with many large steam turbine plants still in service after 30 years of almost continuous operation.

3.3.5 Environmental Considerations

Emissions   As virtually any type of combustible material can be used as the fuel for a steam turbine CHP plant, it is not possible to provide figures for typical exhaust emissions. As has been discussed above, economics usually dictate that steam turbine CHP systems utilise relatively 'dirty' fuels. For this reason, the combustion emissions associated with steam turbine CHP plant are likely to contain significant percentages of noxious gases, along with particulate matter. Large plants may, therefore, require special flue scrubbing equipment to ensure com-

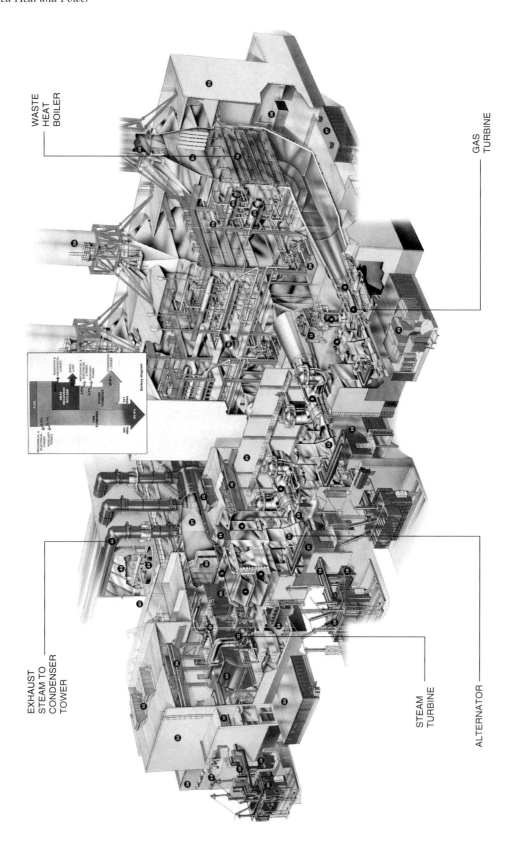

WASTE
HEAT
BOILER

GAS
TURBINE

EXHAUST
STEAM TO
CONDENSER
TOWER

STEAM
TURBINE

ALTERNATOR

Figure 3.3(f) *Exploded View of a Combined Cycle Gas Turbine Power Station*

(Courtesy of Siemens plc)

pliance with environmental regulations. Where refuse and other waste fuels are used, even greater attention needs to be given to the products of combustion and the type of materials that are likely to be burned.

Noise Large steam turbines operating at high speeds generate high levels of noise and therefore need to be enclosed by a suitable building located some distance from residential areas. The noise emissions from boiler burner motors should also not be overlooked, particularly where boilers are located in a separate building to the turbine plant. Noise from smaller turbines can, however, be satisfactorily controlled by suitable acoustic enclosures.

As the working fluid of a steam turbine does not have to be taken from and exhausted to atmosphere but is contained within a closed loop, attenuation of noise at intakes and exhausts is not a consideration. Table 3.3(d) gives the level of noise emitted from the casing of a typical 5 MW steam turbine.

Vibration Like gas turbines, only high speed rotary movement is involved in a steam turbine. For this reason, the transmission of vibration to adjacent structures is not a problem.

Table 3.3(d) *Typical Steam Turbine Noise Emissions*

Source	Noise Emission Levels – dB (A)
Engine Casing[1]	90–100
Intake	n/a
Exhaust	n/a

Notes
[1] Average level measured approximately 1m distatant from a steam turbine with no acoustic enclosure.

3.3.6 Capital and Operating Costs

Capital Costs Like gas turbines, the capital costs of steam turbines with similar power outputs varies markedly from manufacturer to manufacturer, dependent upon the efficiency, maintainability and service life of the machine. To enable broad comparisons to be made between engine technologies, however, typical costs for packaged generating sets based on steam turbines have been provided in table 3.3(c).

Operating Costs The costs that are associated with the operation of steam turbine and boiler plant relate to:

6

HIGHER/LOWER HEAT VALUES AND EFFICIENCY

Gross and Net Energy Content To be able to express the efficiency of any engine it is necessary first to determine the latent chemical energy content of the fuel to be burned. The quantity of energy released by the combustion of a sample of fuel will vary dependent upon whether combustion takes place at constant volume or constant pressure and upon whether the water in the products of combustion is in a liquid or vapour phase. For most hydrocarbon based fuels, the difference in energy release between constant volume and constant pressure combustion is small and hence is usually ignored. The state of the water in the products of combustion, however, can make more than a 10% difference to the measured energy release.

For this reason, when an efficiency figure is given for an engine, it must be clearly stated whether the latent energy content of the fuel used has been taken as 'gross' (water in the products of combustion in liquid phase) or 'net' (water in vapour phase). *It is normal practice for engine efficiency figures to be based on the net latent energy content of the fuel used.* When using the efficiency figures stated by a manufacturer of an engine, therefore, care should be taken to convert calculated input energy figures from net to gross values before fuel mass flow rates and costs are worked out.

Higher and Lower Heat Value In the engine world, it has become common to refer to gross and net latent energy content as 'higher and lower heat value' respectively.

Fuel
Planned maintenance and repairs
Engine and boiler supervision and management

Purely for the purposes of comparison, fuel costs over 10 years have been given in table 3.3(c) using the same nominal gas prices used for table 3.2(b). It should be recognised, however, that the fundamental viability of steam turbines in a CHP application is dependent upon the use of a low cost fuel, making the comparison using a premium fuel such as natural gas a rather spurious one.

Finally, typical costs for planned maintenance and repairs are also presented in table 3.3(c).

3.4 Boilers and Heat Recovery Equipment

3.4.1 Reciprocating IC Engines

Engine Jacket and Oil Cooling System Heat Recovery

Equipment The heat rejected from the engine jacket and lubricating oil cooling systems is recovered to water circuits using conventional shell and tube or plate heat exchangers. Provided the heat recovery circuits themselves can utilise all the recovered heat, virtually 100% of the heat rejected by an IC engine to the jacket cooling water and oil cooling systems can be recovered in practice.

Heat Dumping There may, however, be times when the demand for heat from the heat recovery circuits is less than the rate at which heat has to be rejected by the engine to maintain the engine block and lubricating oil at acceptable temperatures. Under these circumstances, the engine and oil cooling systems must be able to be switched to an alternative means of heat rejection, air blast cooling towers for example. If such provision is not made, the engine will have to be shut down or its power output reduced when demand for recovered heat is less than the required heat rejection rate.

Many small or 'mini' CHP plants do not have alternative engine and oil cooling system heat rejection arrangements. The engines of such plant are, thus, brought on and off line in response to the demand for recovered heat. The omission of alternative heat rejection arrangements does, of course, reduce the capital cost of a project. It should be recognised, however, that such operational constraints may prevent the optimum operation of the plant and hence the achievement of the maximum financial returns.

Typical arrangements for heat recovery and heat dumping from engine and oil cooling systems are illustrated diagrammatically in figures 5.3(a) and 5.3(b) of chapter 5.

Exhaust Gas Heat Recovery

Equipment The relatively low mass flow rate of the exhaust gases from reciprocating IC engines means that heat recovery units can be compact. They are, therefore, usually installed in-line in the exhaust duct, as close as is practically possible to the engine. Fire tube designs are generally used with single or multiple passes to suit the particular application. For large plants the use of water tube designs is, however, not unusual. The units may be used to generate hot water or steam depending upon the use that is to be made of the recovered heat.

A photograph of a vertical exhaust gas heat recovery unit is shown in figure 3.4(a). The unit, fitted to the exhaust of a 1 MW IC engine, is designed to generate steam in conjunction with a steam separation vessel.

Minimum Exhaust Gas Temperature Unlike the recovery of heat from engine and lubricating oil systems, it is not possible in practice to recover 100% of the heat available in the engine exhaust gases. This is due to the requirement to maintain the temperature of the exhaust gases above a certain threshold. Minimum acceptable exhaust temperatures are dependent upon the fuel being burned in the engine and, where the gases are discharged through a tall chimney, buoyancy considerations.

Combustion within the cylinder of an IC engine takes place at extremely high temperatures and hence leads to the formation of combustion products, such as nitrogen oxides, that are not present in the flue gases of conventional boilers. For this reason, particular care must be taken to prevent condensation and the formation of highly corrosive acids in the exhaust gases from IC engines. This is a requirement whether the

(Courtesy of Clayton Thermal Products Ltd)

Figure 3.4(a) *Vertical Exhaust Gas Heat Recovery Unit Serving an IC Engine*

engine is operating on fuel oils or natural gas. The dew point for exhaust gases from IC engines varies with the fuel being burned, ambient air conditions and the loading on the engine at any particular time.

Experience in the US has, however, led to the widespread adoption of a minimum exhaust temperature of 120°C as the design criterion for IC engines. The intentional condensing of water vapour from exhaust gases to achieve greater heat recovery is discussed in section 5.2.1 of chapter 5.

The use of the 120°C design criterion is fine provided the engine is always to be run at or near to design operating conditions. If part load operation is envisaged, however, lower engine exhaust gas mass flow rates will lead to exhaust temperatures of less than 120°C downstream of

the heat recovery unit. For this reason, when it is expected that engine power output will vary it has become normal practice to design for a minimum exhaust temperature of 150°C, thereby providing an adequate margin of safety to prevent condensation during part load operation.

When an engine is being operated on decomposition gases, special materials have to be used for heat recovery units to provide resistance against the highly corrosive products of combustion that are produced.

Back-pressure A second important consideration concerns back-pressure created by pressure drop in the exhaust system. For most IC engines designed for static use and operating at low speed, back-pressures of up to 1.5 kPa are acceptable. For smaller high speed engines a figure of up to 3 kPa may be admissible. Pressure drops in excess of these values lead to loss of power output and a decrease in efficiency of power generation. When back pressure is particularly high, combustion chamber temperatures may increase sufficiently to cause increased valve wear and the associated maintenance problems.

Heat Dumping   In common with the recovery of heat from engine and oil cooling systems, there may be times when the demand for recovered heat is less than the rate at which heat is recovered. If such a situation is likely to arise and the engine is to be kept in operation, then provision has to be made either for the diversion of flue gases around the heat recovery unit or for the dumping of excess heat recovered using an external radiator. Where the heat recovery unit generates steam, the actual rate at which heat is recovered can be reduced by allowing pressure and hence water temperature within the unit to rise, thereby reducing the temperature difference across the heat exchanger surfaces. This technique can, however, only accommodate relatively small differences between demand and heat recovery rate.

Typical IC engine exhaust gas heat recovery arrangements are shown diagrammatically in figures 5.3(a) and 5.3(b) of chapter 5.

Impact on Power Generating Efficiency
Provided maximum pressure drop constraints are met, the heat recovery unit will have little effect upon the performance of the engine. For this reason, the sizing and selection of the heat recovery unit does not have to be undertaken as part of an overall CHP system optimisation process. There is no trade-off between heat recovery performance and efficiency of power generation.

At first glance this appears to fly in the face of the discussion of heat engines in chapter 1, where the relationship between efficiency and input and output heat temperature differential was introduced. Although IC engines are not true heat engines this relationship, nevertheless, still holds. The difference in the case of an IC engine, as opposed to a steam turbine, is that the heat rejection temperatures are fixed at the time of engine design. The CHP system designer

does not, thus, have the option to increase heat rejection temperatures and so is not forced to make decisions on the trade-off between the temperature of recovered heat and the efficiency of power generation. In any case, as table 3.1(a) shows, the exhaust gases from current IC engine designs are at a sufficiently high temperature to satisfy most process heating and all space heating requirements.

The issue of power generating efficiency versus heat rejection temperature is discussed in more detail in insert panel 7 of this chapter.

Supplementary Firing of Exhaust Gases

Where the heat demands at a site exceed the rate at which heat can be recovered from an engine for significant periods of time, supplementary firing of engine exhaust gases is often undertaken. The advantage of supplementary firing is that the exhaust contains some, if not all, the oxygen necessary for the combustion of further fuel. As these gases are already at a relatively high temperature, far less of the heat release by the additional fuel has to be used to heat up air for combustion, in comparison with a conventional boiler.

For complete combustion the exhaust gas must, of course, contain sufficient oxygen. The actual minimum percentage required will depend on the fuel to be used for supplementary firing, the design of the boost burner and the maximum firing rate that is required.

In the case of spark ignition engines operating at stoichiometric air/fuel ratios, exhaust gas oxygen content is less than 1% making supplementary firing a non-starter. Exhaust oxygen content for lean burn spark ignition engines and compression ignition engines, on the other hand, varies from 8% to 12%, making both types of engine potential candidates for supplementary firing. Even for these engines, however, it is usually necessary to add air to the exhaust stream to ensure that adequate oxygen is available for complete combustion of the supplementary fuel.

In practice, the supplementary firing of the exhaust gases from IC engines is not commonly undertaken due to the relatively high cost of the boost burner and forced draught fan systems required. For engines above 1 MW power output supplementary firing can, nevertheless, prove cost effective.

Direct Use of Exhaust Gases
The exhaust gases from an IC engine may be used directly to fire specially designed absorption chillers.

3.4.2 Gas Turbines

Oil Cooling System Heat Recovery
As has been discussed in section 3.2.4, the recovery of heat from the lubricating oil cooling system of a gas turbine is not usually undertaken, as the temperature at which the heat is available is low and the quantity of heat is small in relation to the heat available in the exhaust gases. If an economic case could be made, however, the necessary equipment and arrangements would be similar to those required for IC engines.

Exhaust Gas Heat Recovery

Equipment Heat recovery units on gas turbine exhausts may be used to generate hot water or steam, using both fire tube and water tube boiler designs. As the quantities of heat to be recovered are large, however, it is usual practice to generate steam utilising a water tube waste heat boiler.

Where water tube designs are used, a decision between forced and natural circulation of the water has to be made. Generally, when exhaust temperatures are low (from a gas turbine fitted with a recuperator, for example) and there are physical space constraints, forced circulation is employed. Otherwise, it is normal to utilise a natural circulation boiler design for gas turbine exhaust gas heat recovery. The higher water flow rates that are associated with natural circulation provide two advantages. Firstly, the rate at which salts are deposited onto boiler tube walls is reduced, which results in lower water side cleaning requirements. Then secondly, the more uniform metal temperature that is maintained promotes longer boiler tube life.

A cut-away drawing of a natural circulation waste heat boiler incorporating supplementary firing is shown in figure 3.4(b). A photograph of a similar waste heat boiler arrangement, installed as part of a gas turbine cogeneration system, is shown in figure 3.4(c).

Minimum Exhaust Gas Temperature As with IC engines, combustion in gas turbines takes place at high temperatures and hence leads to the formation of combustion products that are not present in the flue gases of conventional boilers. To eliminate the risk of condensation and hence acid corrosion under varying operating conditions, heat recovery units are usually designed to deliver exhaust gases at a minimum of 150°C under full load operation of the gas turbine.

Back-pressure Whilst the mass flow rate of exhaust gases from a gas turbine considerably exceeds that from an equivalently sized reciprocating IC engine, gas turbines are generally more forgiving of back-pressure than IC engines. It should be recognised, however, that for each 1 kPa increase in exhaust system pressure drop there will be an approximate 1.2% loss in power output from the gas turbine. The design of the heat recovery unit is, therefore, a compromise between pressure drop and engine power output in addition to the normal compromise between heat recovery and capital cost. For the above reasons, heat recovery units for gas turbines tend to be physically large and floor standing.

Impact on Power Generating Efficiency Like IC engines, the temperatures at which heat is rejected from a gas turbine are relatively fixed at the time of design. Once again, therefore, the CHP system designer does not have the option of increasing exhaust gas temperature to suit a particular application. As table 3.2(a) shows, however, gas turbine exhaust gas temperatures are high enough to satisfy most process heating and all space heating applications.

The issue of power generating efficiency versus heat rejection temperature is discussed in more detail in insert panel 7 of this chapter.

Unlike an IC engine driven CHP system, however, the heat recovery unit used in a gas turbine system will have a significant impact on the performance of the engine. For this reason, the design and sizing of the waste heat recovery boiler in a gas turbine CHP plant has to be undertaken as an integral part of the overall system optimisation. For example, increased heat recovery rate or higher heat recovery temperature may have to be traded off against reduced

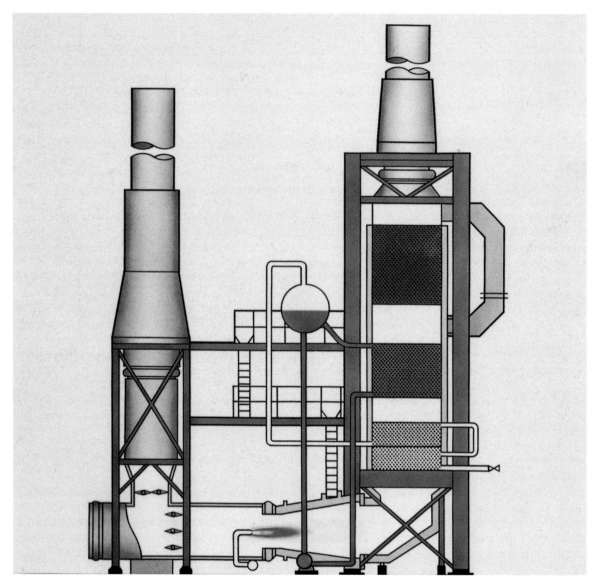

(Courtesy of Senior Thermal Engineering Ltd)

Figure 3.4(b) *Cut-away Drawing of a Natural Circulation Waste Heat Boiler incorporating Supplementary Firing*

power output and efficiency due to the higher back-pressures that will result from the larger heat exchange surface required. In addition, optimum performance from the overall CHP system may require increased air flow rates through the gas turbine and boiler, again reducing power generating efficiency. In a large combined cycle plant for instance, optimum overall power generation performance from the combination of gas turbine and steam turbine

may be achieved with a loss of generating efficiency of up to 2% at the gas turbine, due to increased air flow rates. A typical gas turbine exhaust gas heat recovery arrangement is shown diagrammatically in figure 5.3(c) of chapter 5.

Supplementary Firing of Exhaust Gases
Should higher temperatures or increased heat output be required, fuel can be combusted in the exhaust gases from a gas turbine. The high

(Courtesy of Senior Thermal Engineering Ltd)

Figure 3.4(c) *Two Boiler, Natural Circulation Waste Heat Boiler Installation Under Construction*

excess air utilised in gas turbines provides an oxygen content in the exhaust gases of around 16%, which is sufficient to burn a further 11 kW of fuel per kW of turbine shaft power output. Good mixing of the exhaust gases before supplementary firing is, however, important if complete combustion is to be achieved. As the exhaust gases are effectively being used as a source of highly pre-heated air for combustion, the combustion efficiencies achieved are excellent.

The high oxygen content of the exhaust from gas turbines makes the introduction of additional air for combustion unnecessary. Some boost burner designs do, however, utilise a forced draught fan system to introduce cold air into the exhaust stream around the burner, to cool certain burner components.

Direct Use of Exhaust Gases

Heat recovery boilers are expensive and, due to the back-pressure they create, reduce power generating efficiency. Opportunities to use the exhaust gases directly from a gas turbine should, therefore, not be overlooked. Examples of such applications include the use of exhaust gases in specially designed absorption chillers and the use of exhaust gases to heat and cure items in industrial manufacturing processes.

3.4.3 Steam Turbines

Oil Cooling System Heat Recovery

Like gas turbines, it is unusual for heat to be re-

71

covered from the lubricating oil cooling systems of steam turbines. If an economic case could be made, however, the necessary equipment and arrangements would be similar to those required for IC engines.

Exhaust Gas Heat Recovery

The exhaust gas from a steam turbine is, of course, steam. Exhaust gas heat recovery equipment is not, therefore, required. Where back-pressure turbines are used, exhaust steam pressure and condition will have been chosen to suit the particular heating application concerned. In the case of condensing turbines, steam is exhausted into a condenser with the simple requirement for it to be turned into liquid ready for returning to the boiler. The heat available from the condenser is, of course, at a temperature of less than 100°C and may not, therefore, be of any value in a particular application. As the condensing of the exhaust steam is a requirement of the cycle, however, provision must be made to reject the required quantities of heat from the condenser, whether it is useful or not.

Boilers

The requirements for waste heat boilers raising steam for steam turbines in combined cycle plants, have been discussed in section 3.4.2.

When a steam turbine is to be used as the main engine of a CHP system, boiler design, operating pressures and level of steam superheat are all, of course, critical to the overall performance of the plant. Higher operating pressures and greater steam superheat increase the power generating efficiency of the steam turbine but decrease the combustion efficiency of the boiler plant. Higher steam pressures and temperatures also entail significant capital cost penalties. In common with gas turbines and waste heat recovery boilers, the design of boilers and steam turbines has to be considered together if overall system performance is to be optimised.

The economic use of steam turbines may require the generation of steam at higher pressures and temperatures than would otherwise be required at a site. Under these circumstances, careful consideration should be given to the increased capital and operating costs that may be associated with the operation of the steam system at these more extreme conditions. In cases where increased pressure and temperature to serve a steam turbine is economically justified, the site demand for lower pressure steam should not be met by simply throttling the high pressure steam. A pass-out steam turbine should be used to enable useful work to be extracted from the steam as it is taken from high to medium pressure.

Boiler furnace design is, of course, primarily determined by the fuel to be burned. Water tube type designs are usually utilised.

3.5 Intakes and Exhausts

3.5.1 Reciprocating IC Engines

Intakes

Air has to be drawn from outside into the plant room in which the engine is located for three purposes. Firstly, to remove the heat that is rejected from the outer surfaces of the engine and any associated equipment, exhaust ductwork etc. Secondly, to dilute and remove gaseous fuel released into the space by accident. Then finally, to provide air for combustion.

Combustion Air Depending upon engine type, design and size, 1 to 3 kg/s of outside air per MW of shaft power is required for combustion purposes. As approximately 90% of valve, piston ring and cylinder wall wear is due to dust entrained in intake air, the provision of adequate filtration arrangements and their correct maintenance is important to the achievement of a long service life from these components.

The heating of combustion air is generally avoided as each 10°C rise in temperature reduces shaft power output by approximately 1.5%. Notwithstanding this fact, it is usual for an IC engine to take its combustion air from the engine room rather than directly from outside via ductwork.

Ventilation Air Mechanical ventilation is usually required to ensure that heat emissions within the engine room do not result in excessive room temperatures. In the positioning and selection of the ventilation system intake and extract grilles, care has to be taken to ensure maximum air flow across the engine itself.

7

POWER GENERATING EFFICIENCY VERSUS HEAT REJECTION TEMPERATURE

The concept of Carnot Efficiency, which defines the relationship between power generating efficiency and heat input and output temperature differential, has been introduced in chapter 1. The conclusion drawn from this relationship was that the combination of power with heat generation leads to an inevitable trade-off between power generating efficiency and the temperature at which heat is produced. The discussions of section 3.4, however, have revealed that this trade off may not arise when IC engines and gas turbines are used for cogeneration.

For CHP plants up to 1 MW, in practice, this is in fact the case. In this size range, there are few equipment selection options available to the CHP designer whereby lower heat recovery temperature can be traded for higher power generating efficiency. Above this size, however, two options become practical.

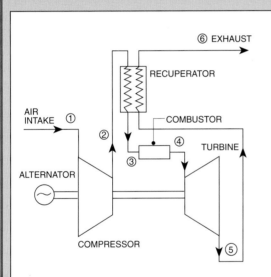

DIAGRAMATIC REPRESENTATION OF GAS TURBINE CYCLE WITH RECUPERATION

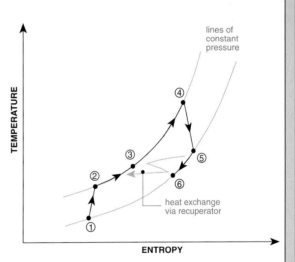

PLOT OF TEMPERATURE AGAINST ENTROPY FOR THE CYCLE

Points in Cycle	Process
① to ②	Air compressed in axial compressor to typically 10 bar; 330°C.
② to ③	Compressed air stream heated at approximately constant pressure to typically 430°C by transfer of heat from exhaust gases.
③ to ④	Fuel combusted in compressed air stream at approximately constant pressure to further raise temperature of the air to typically 970°C.
④ to ⑤	Air and combustion products expanded through turbine down to atmospheric pressure and typically 470°C.
⑤ to ⑥	Exhaust gases cooled at approximately constant pressure to typically 370°C by transfer of heat to compressed air stream.

Figure 3.4(d) *Gas Turbine Cycle with Recuperation*

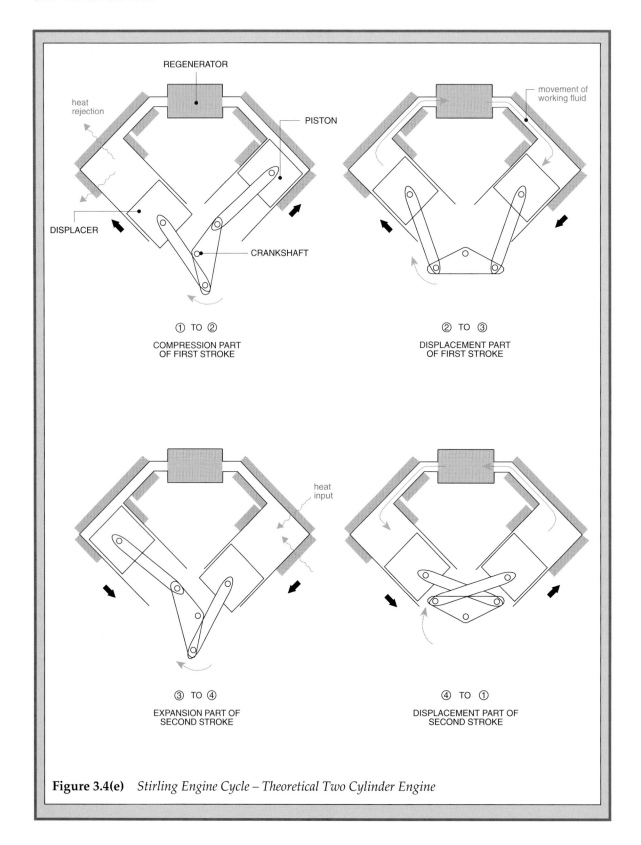

Figure 3.4(e)   *Stirling Engine Cycle – Theoretical Two Cylinder Engine*

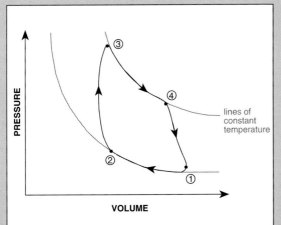

PLOT OF CYLINDER PRESSURE AGAINST VOLUME FOR THE CYCLE

Points in Cycle	Stroke	Operation
① to ②	Compression and	Compression of working fluid at approximately constant temperature with the rejection of heat.
② to ③	Displacement	Displacement of working fluid with recovery of heat from regenerator matrix to further increase pressure.
③ to ④	Expansion and	Expansion of working fluid at approximately constant temperature with heat input.
④ to ①	Displacement	Displacement of working fluid with rejection of heat to regenerator matrix.

Figure 3.4(e) continued *Stirling Engine Cycle – Theoretical Two Cylinder Engine*

Heat Recovery within the Engine Cycle To improve the power generating efficiency of gas turbines, heat from the exhaust gases can be used to pre-heat the air delivered by the compressor of the machine, before it enters the combustion chamber. In this way, less fuel has to be burned in the combustion chamber to achieve the required chamber exit temperature. The necessary arrangements are illustrated diagrammatically in figure 3.4(d). Clearly, with this design of gas turbine plant, lower exhaust gas temperature and hence lower heat recovery temperature is traded for improved power efficiency.

The use of a 'recuperator' to recover heat in this way can improve gas turbine power generating efficiency by more than 5% (i.e. 25% to 30%), whilst exhaust gas exit temperature is lowered, typically, from 500°C to 400°C.

The working fluid in a gas turbine passes continuously through the machine from compressor to combustor to turbine. It is, therefore, possible to transfer heat from the working fluid at the point of exhaust, to the working fluid after compression. It is vitally important for the heat to be transferred after the working fluid is compressed in order that volumetric efficiency of compression and hence power output from the plant is maintained.

In a reciprocating IC engine, however, the working fluid is compressed, fuel is burned and the working fluid is expanded all within the same component of the machine, the cylinder. It is not possible, therefore, to transfer heat from exhaust gases to the working fluid after the working fluid has been compressed.

For this reason, the use of a recuperator on a reciprocating IC engine is not a practical proposition.

There is, however, a cycle around which a reciprocating engine can be designed which is able to make use of exhaust gas heat. The 'Stirling cycle' is detailed in the pressure-volume diagram of figure 3.4(e). Also shown in the figure are diagrams of a theoretical two cylinder 'Stirling engine' to illustrate the operation of the Stirling cycle within a thermodynamic machine.

It will be noted that the Stirling engine is,

in fact, an external combustion engine and therefore operates as a true heat engine. With the recovery of heat from the working fluid following expansion to the fluid following compression, power generating efficiencies are inherently high.

Practical designs of Stirling engines make use of a two piston single cylinder arrangement as shown in the cut-away diagram of figure 3.4(f).

Working fluid used is usually helium at high pressure and, being an external combustion engine, almost any combustible material can be used to generate the heat needed to drive the cycle. With the additional benefits of relatively simple mechanical components and quiet operation the Stirling engine appears, at first sight, to be the perfect plant to drive a cogeneration system.

There are, however, significant materials technology difficulties in producing a regenerator that is both effective and that can operate continuously at the high temperatures required. For this reason, though a number of wood fired stirling engines were built at the turn of the century, few working machines have been developed recently. A number of highly successful low temperature refrigeration plants have, nevertheless, been developed which utilise the stirling cycle. A low level of development work on the heat engine technology continues and it appears that commercial Stirling engines may reach the market place within the next 5 to 10 years.

Combined Cycle The second option is to raise steam using the exhaust gases of either an IC engine or a gas turbine and use that steam to drive a steam turbine and hence generate further power i.e. 'combined cycle'.

Gas turbines are most commonly used as the primary engine of combined cycle plants and this type of arrangement has been discussed in detail in insert panel 5 of this chapter. The comparatively low proportion of rejected heat that is available at a high temperature in the exhaust gases from an IC engine, makes combined cycle uneconomic for most installations that use IC engines. The exception to this generalisation are large plants, which characteristically reject a higher proportion of their heat from the engine exhaust. Combined cycle systems that recover heat from marine diesel engines on board ships, for example, are not uncommon.

From the above, it can be concluded that for medium to large scale installations the decision to use CHP does entail a trade off between heat recovery and power generating efficiency. For installations up to 1 MWe, however, the recovery of heat from the engine plants available in this size range has little or no impact on power efficiency.

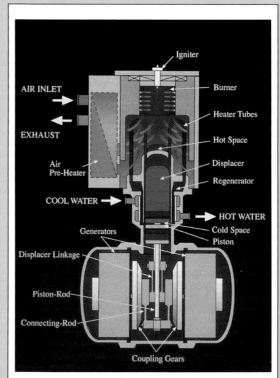

(Courtesy of EA Technology Ltd)

Figure 3.4(f) *Cut-away Diagram of a Two Piston, Single Cylinder Stirling Engine*

Where an engine is used intermittently, motorised dampers are required to close intake and exhaust openings when the engine is not running, to ensure that adequate engine room temperatures are maintained.

Ventilation air rates chosen to satisfy heat removal criteria, are generally also satisfactory for the purpose of dilution and removal of gases from fuel leaks.

Exhausts

Both air used for ventilation and the products of combustion from the engine need to be exhausted from the engine room.

Combustion Products   These have, of course, to be ducted directly from the exhaust manifold on the engine. Flexible connections are used to minimise the transmission of vibration. These are selected to be able to withstand the peak exhaust gas temperatures for the chosen engine, which may be as high as 750°C for short periods of time with gas driven engines.

Even though exhaust gas temperatures are maintained above dew point under normal operating conditions, drain points will be installed at low points to remove the condensate that forms on machine start up. In addition, horizontal runs are always installed with a slight fall away from the engine to eliminate the risk of condensate flowing back into the engine itself.

Individual exhausts are provided for each engine to reduce the risk of condensation in the engine and silencer of off-line machines.

As has already been mentioned in section 3.4.1, the overall pressure drop through the exhaust ductwork, silencer and the waste heat recovery unit must be kept to approximately 1.5 kPa. The actual length of the exhaust pipe can, however, also have a significant impact on the performance of IC engines, as a result of the pressure pulses and reflections that are set up within the exhaust system. Specific advice is, therefore, needed from the manufacturer of the chosen engine to ensure that exhaust pipe length is correctly selected to optimise engine performance.

To reduce the quantity of ventilation air required for engine room cooling, all silencers, heat recovery units and exhaust ductwork

within the room are clad with high temperature insulation.

3.5.2 Gas Turbines

Intakes

In common with IC engines, outside air is required for three purposes: to remove heat rejected from the outer surfaces of the turbine, to dilute and remove accidental releases of gaseous fuel and to provide air for the engine.

Combustion Air/Working Fluid   In the reciprocating IC engine, the mass flow rate of working fluid that produces optimum performance happens to be almost exactly the same as the mass flow rate of air required for complete combustion of the fuel. In other words, near stoichiometric ratios of air to fuel are required. In a gas turbine, however, the completely different cycle processes involved mean that optimum performance is achieved at far greater working fluid mass flow rates. Mass flow rates of air significantly in excess of those required for complete combustion are, therefore, needed in gas turbines.

Between 4 and 8 kg/s of outside air per MW of shaft power is required by gas turbines, in contrast to the 1 to 3 kg/s per MW for IC engines. This higher ratio of working fluid mass flow rate to power output makes gas turbines particularly sensitive to changes in input air conditions. A 10°C rise in ambient air temperature, for example, will decrease power output by 6 to 10% depending upon machine design. The equivalent figure for IC engines is 1.5%.

Due to the high mass flow rates required and the need to use the coldest available air, ambient air is usually ducted directly from outside onto the air intake of the gas turbine. The high levels of noise generated by the machine, however, mean that sound attenuation is required in the intake duct in almost all situations. This problem can be mitigated by utilising a vertical air intake where there are no adjacent tall buildings.

Unfortunately, pressure drop in the intake ductwork significantly affects performance. Each 1 kPa of pressure drop reduces power output by roughly 2%. For this reason, sound attenuators have to be carefully selected to achieve the required level of attenuation and no more, to keep pressure drop to a minimum.

An adequate level of intake air filtration is critical to the maintenance of power generating efficiency for gas turbines. The issue of intake air filtration is discussed in section 5.2.3 of chapter 5.

Ventilation Air   The requirements for gas turbines are generally the same as for IC engines. Where machines are installed within sound attenuating enclosures, separate ventilation arrangements are provided for the enclosure.

Exhausts

Both air used for ventilation and gas turbine combustion products have to be exhausted from the turbine hall.

Combustion Products   The products of combustion exhausted from the gas turbine are ducted directly to the waste heat boiler and then away from the boiler using a conventional flue. A bypass ductwork arrangement is usually provided to divert the exhaust gases around the boiler at times of low heat demand or when the boiler is down for maintenance. As vibration from gas turbines is not a significant problem, flexible connections are not technically required. For practical installation convenience they are, nevertheless, usually fitted. In common with IC engines, good designs incorporate drainage arrangements and falls on horizontal ductwork runs to prevent condensate, formed during start up, running back to the turbine.

As discussed in section 3.4.2, back-pressure significantly affects the power output from gas turbines. Exhaust systems, heat recovery boilers and any additional sound attenuation arrangements required are, thus, designed to keep pressure loss to a minimum. To reduce the quantity of ventilation air required for turbine hall cooling, exhaust ductwork, heat recovery boilers and flues are clad with high temperature insulation.

3.5.3 Steam Turbines

Intakes

Once again, outside air is required for three purposes: to remove heat rejected from the outer surfaces of the turbine, to dilute and remove ac-cidental releases of gaseous fuels and to provide air for combustion in the steam raising plant. In the case of the steam turbine, however, the working fluid used in the engine is steam not air.

Combustion Air   The fuel for a steam turbine system is burned in a steam boiler. The requirements for combustion air are, therefore, the same as for any conventional boiler utilising the chosen fuel.

Working Fluid   As mentioned above, the working fluid is steam at high pressure and temperature raised by the boiler plant. Steam is transported to the 'steam chests' of the turbine using conventional steel pipework of sufficient weight to withstand the maximum system pressures expected. Noise emissions from the intake to steam turbines are not, therefore, a problem.

A motorised control valve or series of valves are installed in the steam supply to provide power output control for the turbine.

Ventilation Air   The requirements are generally the same as for IC engines and gas turbines.

Exhausts

In a steam turbine system, ventilation air has to be exhausted to atmosphere as do the products of combustion from the boiler plant. The steam used by the turbine, however, is either used for heating in the case of back-pressure and pass-out turbines or is simply condensed and returned to the boiler in the case of condensing turbines.

Combustion Products   The requirements for the exhausting of combustion products are, of course, exactly the same as for any boiler operating on the chosen fuel. As mentioned in section 3.3.5, however, on large plants the products of combustion may require treatment before they can be released into the atmosphere, if particularly dirty fuels are being burned.

3.6 Gas Compressors

3.6.1 Reciprocating IC Engines

In spark ignition IC engines, the air/fuel mix is drawn into the cylinders by the sub-atmospheric pressure created by the descending pistons. Where gas is used as the fuel the supply pressures required are normally minimal and district pressures are usually adequate. Under these circumstances no gas compression equipment is required. Where turbo-charged spark ignition engines are to be operated on gas, machines that introduce the gas at the inlet to the turbo-charger are usually selected to avoid the need to provide a high pressure supply.

Finally, in the case of dual fuel engines, extremely high fuel gas supply pressures are required. The discussion on gas compression given in section 3.6.2, in relation to gas turbines, therefore also applies to dual fuel IC engines.

3.6.2 Gas Turbines

As has been discussed in section 3.2.3, gas turbines require a high pressure gas supply, typically between 12 and 20 bar. This represents an 11 to 18 fold increase over normal district supply pressures. With volume rates approaching 500 l/s for a 5 MW gas turbine, the fuel gas compression installation required for a CHP plant is substantial.

Method of Compression For gas turbine installations it is normal practice to use either reciprocating or double screw machines for fuel gas compression.

Screw compressors, as a result of the rotary compression action employed, produce minimal pressure fluctuations at the inlet and outlet of the machine. Reciprocating machines, in contrast, generate regular pressure pulses which can resonate in a gas supply system to create quite violent pressure waves. The associated problems are discussed in section 5.3.3 of chapter 5. This does not, however, preclude the use of reciprocating machines, as the pressure pulses produced can be reduced to acceptable levels through the judicious engineering of the overall supply system.

Efficiency The full load performance of a reciprocating compressor is generally superior to that of an equivalent screw machine. When it comes to efficiency at part load, however, screw compressors have the edge.

Whilst the economics of cogeneration dictate that high engine loadings are maintained for the majority of the year, fuel gas compression plant will be required to operate part loaded for the following reasons. Firstly, the maximum power output from a gas turbine (and hence demand for fuel gas) may vary by more than 25% during the year, as a result of changing ambient temperature. Secondly, in the case of reciprocating machines, the maximum output from the compressor reduces with service, making it normal practice to install compressors with an output when new which is 15% to 25% greater than the maximum demand for gas. Part load compressor performance is, thus, an important consideration where compression plants for gas turbines are concerned.

Control Due to the rates at which fuel is consumed by a gas turbine, it is normal to store only sufficient compressed gas in the receiver vessel to supply maximum gas demand for 10 seconds, or so. For this reason, simple on/off or step control of compression plant will not work.

The normal form of control used on reciprocating machines, namely cylinder unloading by the lifting of suction valves, is, thus, not satisfactory for the gas turbine application. The problem can be overcome though the use of a modulating bypass valve to control outlet pressure at a constant value by diverting compressed gas back to the intake of the machine. The disadvantages of this approach include poor part load performance and the requirement for bypass gas cooling arrangements.

A second technique on reciprocating compressors is to employ variable compressor speed using inverter drive motor control. Though more expensive than bypass control in capital terms, variable compressor speed control gives good part load efficiency. In the design of the control strategy for this technique, care must be taken to avoid the possibility of hunting at the point of changeover from speed control to suction valve off-loading. In addition, at low operating speeds, compressors may require enhanced means of lubrication.

Finally, a compromise solution to the problem of providing modulating control for reciprocating compressors, in terms of capital cost versus efficiency, is to use 'stepless unloading'. This form of control achieves flow modulation by varying the degree to which suction valves are held open, thereby reducing the effectiveness of each piston stroke in a stepless fashion.

(Courtesy of Peter Brotherhood Ltd)

Figure 3.6(a) *Packaged Fuel Gas Compression Plant – Screw Compressor*

(Courtesy of George Meller Ltd)

Figure 3.6(b)   *Packaged Fuel Gas Compression Plant – Reciprocating Compressor*

Screw machines, on the other hand, naturally provide modulating control of volume flow rate through the use of an axial slide valve which acts to reduce the effective displacement volume of the compressor. The result is that reasonable compression efficiency is maintained down to, typically, 15% loading. Below this level, either suction throttling or gas bypass must be employed, with the consequent loss of efficiency.

Maintenance, Size and Vibration   Screw compressors have three further advantages over reciprocating machines. Firstly, the rotary action employed in a screw machine demands less moving parts than are required in an equivalent reciprocating machine. This gives screw compressors the advantage of higher reliability and lower maintenance demands. The second advantage concerns size. A screw compressor of a given capacity will generally be more compact than an equivalent reciprocating machine. The final advantage relates to vibration. Screw compressors, due to their inherently balanced rotary

action, generate a lower level of vibration compared to reciprocating machines. The consequence is that foundations and vibration isolation requirements are less onerous for screw compressors.

Photographs of two packaged skid mounted gas supply plants, one utilising a screw compressor and the other a reciprocating compressor, are shown in figures 3.6(a) and 3.6(b) respectively.

Oil Carry-over   The carry-over of lubricating oil from fuel gas compressors can have an impact on the rate of combustor and turbine blade fouling. For this reason, fuel gas compression plants incorporate oil separation vessels at the compressor discharge to reduce oil carry-over to, typically, less than 5 ppm (by mass) for a gas turbine application.

Both reciprocating and screw designs of compressor are manufactured in 'dry' versions. Such designs use low friction materials at points of contact within the compression spaces, thereby negating the requirement for oil lubrication within the volumes through which the fuel gas passes. In this way, dry machines are able to produce compressed gases with zero lubricating oil content. For certain processes this is important. In the case of gas turbines, however, the total eradication of oil carry-over is not necessary. It is, therefore, unusual to find a dry design of compressor serving the gas turbine of a CHP plant.

3.6.3 Steam Turbines

The gas supply for steam turbine CHP plant is required for conventional steam boilers. Pressure requirements are, therefore, the same as for standard boiler burners.

3.7 Alternators

In this book the word 'alternator' has been used throughout to refer to 'ac generators'. This has been done to eliminate the potential for confusion with the common use of the term generator to describe a complete generating set (engine and alternator).

3.7.1 Asynchronous

On small generating sets, typically 100 kWe or smaller, the use of asynchronous alternators is an option. In an asynchronous machine, the magnetising current required in the windings of the rotor (excitation current) is induced by the rotating magnetic field produced by the main current flowing in the surrounding stator windings. Hence the alternative term 'induction generator'. A generating set fitted with an asynchronous alternator is run up to starting speed by connecting the stator windings of the alternator to an external electrical supply, thereby driving the alternator as an induction motor. Once the engine of the set fires, the rotor of the alternator will continue to accelerate up to synchronous speed, at which time the alternator will neither draw nor generate power. Once synchronous speed is exceeded, the alternator will start to generate power.

There are several advantages to this method of running up a generating set. Firstly, as the external electrical supply is used to start the set, the output from the generator is inherently synchronised with the external supply. No special

synchronising control and protection equipment is, thus, needed. Secondly, no starter motor is required to motor the engine of the set up to starting speed. The final advantage concerns fault current. In the event of a fault on the supply, an asynchronous alternator will not contribute several times its nominal rating to fault current, in the same way that a synchronous alternator does. As the magnetic flux in the rotor decays, however, an asynchronous alternator will make a fault contribution, all be it a modest one, in the same way that induction motors contribute fault current, in the event of a supply failure.

The major disadvantage of asynchronous alternators concerns the lack of flexibility in their operation. As excitation current is induced, it cannot be easily controlled to regulate the output voltage or power factor from the alternator to compensate for variations at the site. Indeed, asynchronous alternators generate at a leading power factor which requires some form of reactive compensation, if it is not to place a lagging reactive load on the system.

A potential second disadvantage is self-evident. As an asynchronous alternator requires an external electrical supply to induce rotor excitation current, such a machine can not be operated to provide standby power in the event of a total supply failure. It would be unusual, however, to either size a CHP plant to meet the maximum electrical load at a site or provide load shedding facilities, at the size of site for which the employment of asynchronous alternators is likely to be considered. For this reason, the use of a small CHP plant on its own to provide standby power is likely to be precluded due to capacity constraints, making the question of external supply excitation irrelevant.

3.7.2 Synchronous

For generating sets above 100 kWe, synchronous alternators are usually used. Synchronous machines incorporate auxiliary direct current generators (on the same shaft as the alternator) which produce rotor excitation current for the alternator. Such machines can, therefore, be operated in the absence of an external electrical supply.

Smaller synchronous alternators and engines may be started in a similar fashion to asynchronous generators, using an external supply

8

GENERATOR CONTROL

Ignoring auxiliary systems, generator control is essentially composed of two components:

- Engine speed control
- Alternator excitation control

The form of control used in each case is discussed in the paragraphs which follow.

Engine Speed Control   The fuel supply to an engine is generally controlled by an automatic governor with reference to engine speed. Under 'droop control', the control of fuel is set to run from minimum (zero engine load) at design engine speed to maximum (full engine load) at, typically, 95% of design engine speed. This is termed '5% droop'. Droop control is simple 'proportional only' control.

An 'isochronous' governor controls engine speed using 'proportional plus integral action' in the control strategy to maintain design speed from zero to full loading. Engine loading is, of course, determined by the electrical power delivered by the driven alternator.

Alternator Excitation Control   A synchronous alternator is designed to produce design output voltage when supplying rated current, using a nominal rotor excitation current. As load on the alternator falls and hence current supplied reduces, output voltage will increase. This is illustrated by the unity power factor plot drawn on the graph in figure 3.7(b). The difference between output voltages at design and zero current supplied as a percentage of output voltage at design current is termed the 'percentage voltage regulation' for the alternator.

When an inductive load is placed on an alternator, current will lag voltage which causes a reduction in voltage generated in the alternator windings for a given rotor excitation current. The converse is true for capacitive loading. The impact of reactive loading on the relationship between output voltage and current supplied is also shown in figure 3.7(b) –

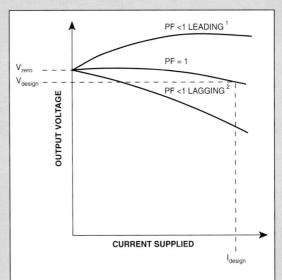

OUTPUT VOLTAGE AGAINST CURRENT
SUPPLIED FOR A SYNCHRONOUS ALTERNATOR

$$Percentage\ Voltage\ Regulation = \frac{V_{zero} - V_{design}}{V_{design}} \times 100$$

where

I_{design} – Design Current Rating
V_{design} – Voltage at Design Current Rating
V_{zero} – Voltage at Zero Current
PF – Power Factor

Notes
1. Leading power factor caused by capacitive loading.
2. Lagging power factor caused by inductive loading.

Figure 3.7(b)   *Voltage Regulation of an Alternator.*

inductive and capacitive loading producing lagging and leading power factors respectively. When giving the percentage voltage regulation for an alternator it is, therefore, necessary to state the power factor on which the figure is based (usually unity or 0.8 lagging).

For an alternator to maintain a constant output voltage at different supply power factors it is, thus, necessary to vary rotor excitation current. As higher excitation current produces higher output voltage, to combat the increased voltage regulation caused by a lagging power factor, excitation current must be increased. The control of rotor excitation current is known as 'automatic voltage regulation (AVR)' control.

Where a site is operating in island mode, the power delivered by the alternator of the local generating set will be determined by the load on the system. Under these circumstances, AVR control will be set to maintain the required voltage at the paralleling bus-bars.

When paralleled with an external supply, generator 'sending end voltage' will determine the level of import to or export from the site from/to the utility company network (sending end voltage is discussed in section 3.7.4). It is, therefore, generator AVR control which determines the level of power delivered by an alternator operating in parallel mode.

An interesting control situation arises when a CHP plant is to be controlled under a 'heat led' strategy (see chapter 8, section 8.1.1). Under such a strategy, it is the heat output from the engine that is to be controlled to match the heat load at the site. Engine heat output is, however, determined by engine power output which cannot be controlled directly but is, in turn, determined by alternator power delivery. An increase in heat output from a cogeneration system in response to an increase in site heat demand, therefore, actually requires an increase in alternator excitation under generator AVR control. Should a site become islanded and the CHP plant be required to provide standby power, then plant control would have to switch to an 'electrical led' strategy immediately, to ensure that site electrical demand was satisfied.

Finally, some utility company transformers are fitted with their own 'automatic voltage control (AVC)'. Where local generating plant is to be operated in parallel with an external supply that incorporates AVC at local transformers, then generator AVR control, when set to maintain paralleling bus-bar voltage, may fight against the transformer voltage control. Under these circumstances, it is recommended that AVR control is switched to maintain a set power factor rather than a set voltage. On a return to island mode, AVR control would need to be switched back to voltage control.

directly to drive the alternator, as an induction motor, up to the starting speed of the engine. It is also possible to start larger sets in this fashion using a star/delta starter or an inverter drive soft start unit. For CHP plants, however, the use of a separate starter motor is usual.

Where a synchronous generator is started using either an inverter drive or a starter motor, the alternator of the set must be run up to speed disconnected from any external supply. When the generating set has reached design speed, the output from the alternator must then be synchronised with the external supply before the alternator can be connected to the supply system.

This requirement for synchronisation demands the provision of sophisticated automatic controls to achieve and maintain synchronism. In addition, special protective measures are required to ensure that an alternator which can no longer maintain synchronism is disconnected from the distribution system and that an unsynchronised alternator can not inadvertently be connected to the system.

Generating sets equipped with synchronous alternators may add significantly to the prospective fault level on both the site distribution system and the external supply network. Where excitation control is automatic, typically a synchronous machine will contribute 3 times its rated output under 3-phase fault conditions. The implications of increased fault levels are discussed in section 5.4.3 of chapter 5.

As rotor excitation current is provided externally, rather than induced as in an asynchronous machine, rotor excitation can be closely controlled to vary the voltage and hence power output and power factor from the alternator. A synchronous alternator can, therefore, be controlled to boost voltage at a site when external supply voltage is depressed or improve power factor. For larger installations this type of operating flexibility can be important.

A cut-away drawing of a synchronous alternator is shown in figure 3.7(a).

3.7.3 Heat Rejection

The very largest of alternators used in power stations are cooled by hydrogen, circulated by a dedicated heat rejection system. Hydrogen is used due to its superior heat rejection properties in this application. In CHP installations and modern power stations up to typically one hundred megawatts output, air cooled alternators are used.

The alternators themselves contain the necessary fan blades to draw air from the engine room through the casing of the unit. No dedicated mechanical ventilation systems are, therefore, required. Where the complete generating set is installed inside a sound attenuating enclosure, ventilation air is supplied to the alternator end and exhausted from the engine end of the enclosure to ensure that cool air passes through the alternator.

3.7.4 Generating Voltage

For CHP installations the choice is between high voltage at, typically, 11 kV and low voltage. Generation at high voltage is more efficient as alternator winding losses are lower due to the smaller currents needed per kW of electrical output.

The switching and protection requirements associated with electrical generation, on the other hand, are far more onerous and expensive at high voltage than at low voltage. The balance of economics means that it is rarely cost effective to generate at high voltage below approximately 1 MWe.

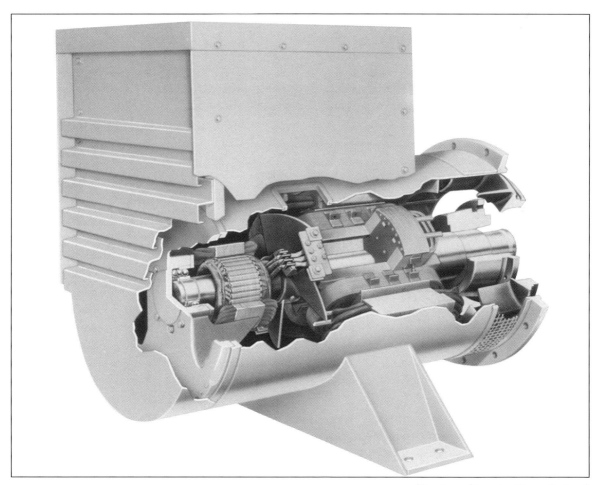

(Courtesy of Leverton Power Systems)

Figure 3.7(a)   *Cut-away Drawing of an Alternator*

Figure 3.8(a) *Simplified Line Diagram of Typical Connection and Protection Arrangements for Low Voltage Systems with Embedded Generation.*

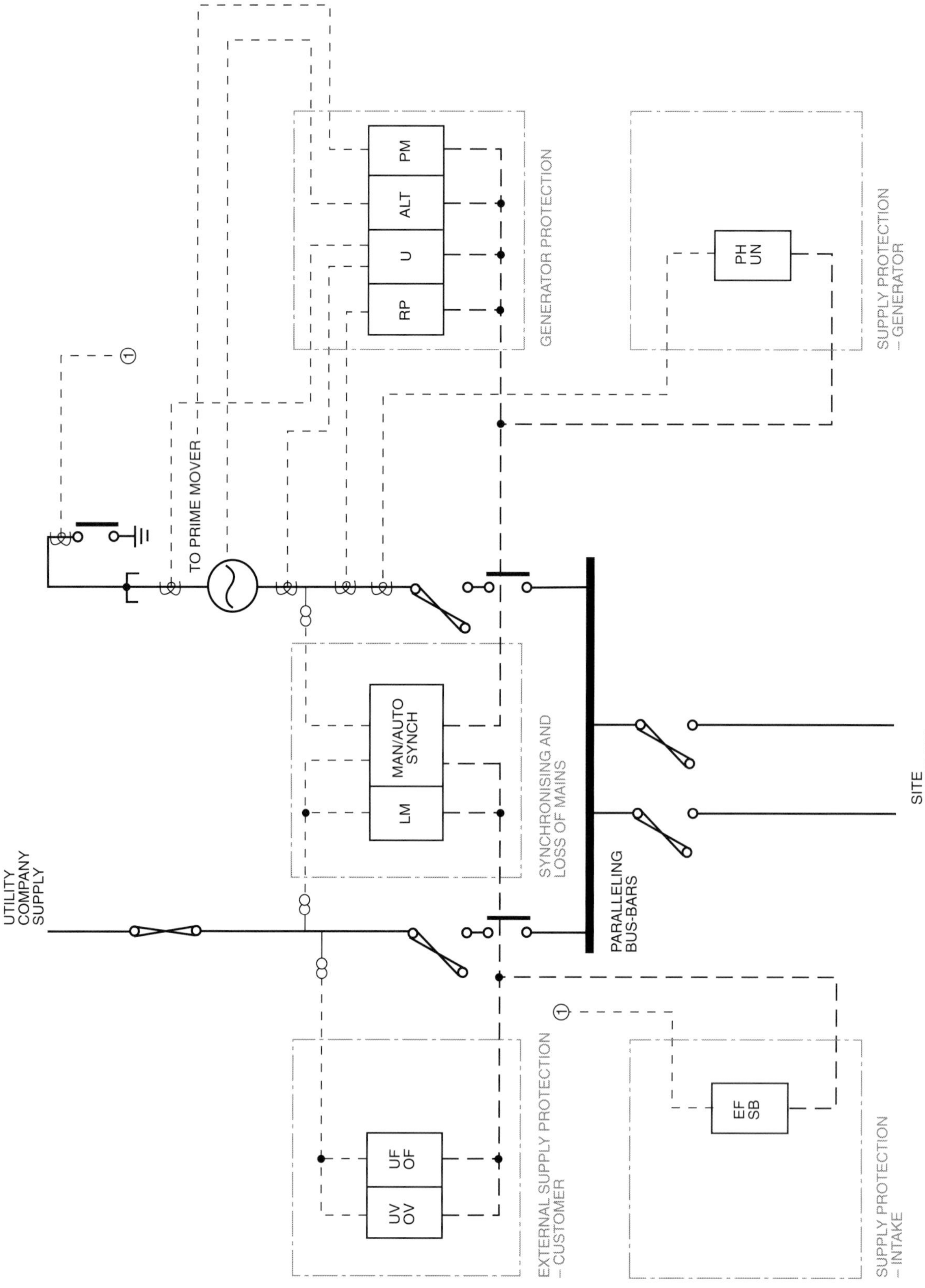

KEY

| MAN/AUTO SYNCH | MANUAL AND AUTOMATIC SYNCHRONISING EQUIPMENT |

| UV OV | PROTECTION EQUIPMENT: |

ALT – ALTERNATOR
EFSB – STANDBY EARTH FAULT
LM – LOSS OF MAINS
PHUN – PHASE UNBALANCE
PM – PRIME MOVER
RP – REVERSE POWER
U – UNIT
UFOF – UNDER/OVER FREQUENCY
UVOV – UNDER/OVER VOLTAGE

FUSESWITCH

CONTACTOR

EARTH ELECTRODE

CURRENT TRANSFORMER

VOLTAGE TRANSFORMER

Figure 3.8(b) *Simplified Line Diagram of Typical Connection and Protection Arrangements for High Voltage Systems with Embedded Generation.*

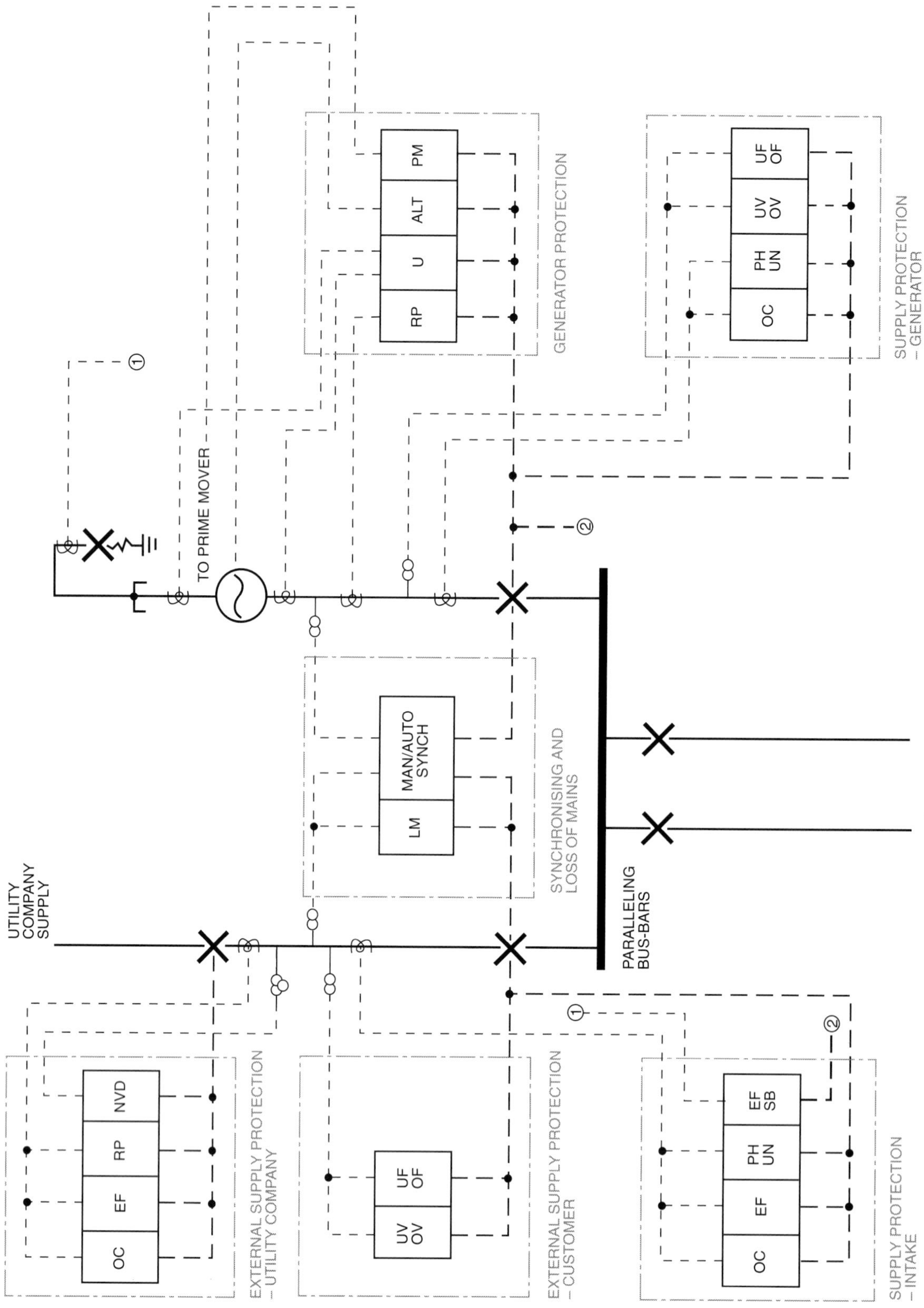

KEY

| MAN/AUTO SYNCH | MANUAL AND AUTOMATIC SYNCHRONISING EQUIPMENT |

| UV OV |

PROTECTION EQUIPMENT:

ALT – ALTERNATOR

EF – EARTH FAULT

EFSB – STANDBY EARTH FAULT

LM – LOSS OF MAINS

NVD – NEUTRAL VOLTAGE DISPLACEMENT

OC – OVERCURRENT

PHUN – PHASE UNBALANCE

PM – PRIME MOVER

RP – REVERSE POWER

U – UNIT

UFOF – UNDER/OVER FREQUENCY

UVOV – UNDER/OVER VOLTAGE

CIRCUIT BREAKER

EARTH RESISTOR

EARTH ELECTRODE

CURRENT TRANSFORMER

VOLTAGE TRANSFORMER

A second consideration regarding generating voltage concerns the export of power. Where a CHP plant is to export power to a local utility company, it will be necessary to generate at a sufficiently high voltage to overcome the impedance of the company's feeder circuit, if electricity is actually to flow *to* rather than *from* the utility. Where feeder circuit impedance is high, the 'sending end voltage' required at the alternator terminals may exceed limits set down by local codes and regulations and might cause other utility company customers to experience excessive voltages. Under such circumstances, a second parallel feeder circuit would need to be installed to reduce the supply impedance.

3.8 Electrical Switchgear and Protection

To be considered effective at a site which has its own generating plant, switching and protection arrangements must:

a) Clear faults with sufficient speed to: limit the touch hazard presented to humans when extraneous metal parts become live due to a fault; limit damage at the point of the fault; prevent damage to conductors and switchgear and prevent damage to generating plant.

b) Meet the requirements of the utility company with regard to the protection of other local customers from unacceptable output from the site generating plant, in the event of the loss of network supply.

c) Achieve adequate discrimination between the operation of protective devices throughout the site to ensure that healthy circuits remain connected in the event of a fault.

d) Be designed to maximise the security with which the site is provided with an electrical supply.

As discussed in section 3.7.2, local or 'embedded' generating plant will add to the prospective fault level on a supply. The conductors and switchgear of both the site distribution system and the external supply network must be capable of withstanding the combined fault level. The issue of reinforcement is discussed in section 5.4.3 of chapter 5.

Simplified diagrams of typical connection and protection arrangements for low and high voltage systems with embedded generation are given in figures 3.8(a) and 3.8(b).

3.8.1 Operating Mode and Fault Levels

Island Where a site is geographically remote or at times when an external electrical supply is not available, a generating set may be operated in 'island mode' i.e. physically disconnected from an outside electricity network. Under these circumstances, generating frequency has to be kept at 50 or 60 Hz to meet the requirements of equipment at the site but no synchronisation with an external supply is, of course, required. Where the site is served by more than one generating set, however, the output from the sets has to be synchronised.

Parallel In general, the output from a CHP installation is not sized to meet the maximum demand for electricity at the site served. A permanent connection must, therefore, be maintained with an external supply, to enable electricity to be imported onto the site to meet demands in excess of cogeneration plant capacity. CHP generating sets are, thus, operated in parallel with the external supply. In 'parallel mode', the 3-phase output from each generator must be synchronised with the external supply. Synchronisation is discussed in section 3.8.3.

Fault Levels Where a site, equipped with generating plant, is designed to be operated both with and without a connection to an external supply, the prospective fault level at any point on the distribution system may differ by several orders of magnitude depending on whether the site is paralleled or islanded.

Consider a site which is connected at high voltage to the supply network of a utility company and also has a 1 MVA generator sized to support the entire site electrical load. With the site connected to the external supply, a 3-phase fault at the site paralleling bus-bars might result in a fault level of several hundred MVA. In contrast, should the same fault occur when the site is islanded, a fault level in the region of just 3 MVA would, typically, be provided by the on-site generator.

It will be realised, therefore, that the protection arrangements provided at a site which is to operate in both parallel and island modes have to be designed to provide effective protection over a wide range of fault currents.

3.8.2 Physical Connection Arrangements

Point of Connection   At most sites, it is convenient and cost effective to connect the alternator of a CHP plant as close as is practicably possible to the point of electrical intake to the site.

Where a high voltage ring exists at a site, it may be possible to connect the alternator at some point on the ring other than at the main intake. This is sometimes desirable where the CHP plant has to be located a significant distance from the point of intake. One option, particularly for alternators in the 1 MWe size range, is to generate at low voltage and transform the output up to high voltage for distribution, using a delta/star step-up transformer. Where non-standard arrangements such as these are used, particular care has to be taken in the design of such arrangements to ensure that adequate protection is provided under all conceivable fault conditions.

In the choice of physical location for the CHP plant, connections to existing heating circuits and/or steam mains often prove the overriding consideration. This is simply due to the fact that high voltage electricity can be transported at lower cost than hot water or steam (see table 2.3(a) of chapter 2). At many sites, therefore, it will have been necessary to run a cable some distance from the CHP plant to the main electrical intake board.

3.8.3 Generator Protection and Synchronising

The purpose of 'Generator Protection' is to prevent damage to generating plant in the event of engine and alternator failures or electrical faults on the plant and the supply system.

Prime Mover Protection   'Prime Mover Protection' is comprised of two components. The first concerns the prevention of unnecessary damage to the engine should a mechanical or control failure occur on the engine itself. This is achieved through the monitoring of various engine operating parameters such as oil pressure, coolant temperature, bearing vibration and speed. In the event that one of these parameters falls outside pre-set limits, the fuel supply to the engine is immediately shut-off with the intention that the engine is brought rapidly to rest, thereby avoiding any further mechanical damage.

The second component of Prime Mover Protection is related to the requirement to bring an engine to a stop in as short a time as possible following engine failure. Should a generator remain connected to an external supply when a prime mover failure occurs, then the alternator of the set will be driven as an induction motor, with the result that the engine can not be brought to rest rapidly. To ensure that 'generator motoring' does not occur, the detection of a prime mover failure will directly cause the generator breaker to trip, in addition to shutting off the engine fuel supply.

Generator Reverse Power Protection   For undetected problems that lead to engine failure, automatic generator breaker trip is provided by a relay which detects the reversal of power at the alternator output terminals (as this indicates that the alternator is no longer generating but is being driven).

Generator Unit Protection   Protection of generating plant in the event of a short circuit or earth fault on an alternator is provided by 'unit protection relays'. These relays, which are connected to current transformers located on the phase and neutral conductors from the alternator, are arranged to trip the associated generator breaker immediately should an alternator short circuit or earth fault occur. 'Unit Protection' can be installed to detect faults on feeder units, transformers and reactors as required.

Unit Protection is backed-up by 'overcurrent and earth fault relays', which form part of 'Site Supply Protection'. Site Supply Protection is discussed in sections 3.8.4 and 3.8.5.

Alternator Overheat Protection   Protection of alternators against overheating due to overload, fault current or current unbalance is discussed elsewhere in this section. Alternator overheating can, however, occur for other reasons such as internal winding failures, mechanical failures and the loss of the ventilation air supply for cooling. To prevent serious alternator damage under these conditions, sensors embedded in the windings of the alternator are used to shut the generating set down and trip the associated generator breaker if high temperatures are detected.

Alternator MVAr Overload Protection The possibility of 'reactive power overloading' of alternators is eliminated through the use of 'over-excitation limiters' on the controls which regulate alternator excitation current. When a site is islanded, however, it is important that over-excitation of alternators is allowed for a short period of time, to enable generators to contribute effectively to the clearance of faults (by generating several times their rated output). For this reason, over-excitation limiters incorporate an inverse time characteristic.

This form of protection, known as 'MVAr Overload Protection', is, of course, only required where synchronous alternators are used.

Alternator Excitation Failure Protection Another potential failure, where synchronous machines are concerned, is the complete loss of excitation to the alternator during parallel operation. Under these circumstances, the alternator acts as an asynchronous machine with the result that rotor speed rises until the new speed/torque characteristic for the alternator matches the engine governor setting. With alternator rotor speed exceeding synchronous speed by a significant margin, local voltage will be depressed and excessive heating of the rotor may occur.

To ensure that such a situation is not allowed to persist, 'Loss of Field Protection' is required. This is usually achieved through the use of 're-verse reactive current relays' which are designed to trip the generator breaker on detection of loss of field, after a set time delay.

Loss of Mains Protection On the loss of the external supply, where local generating plant continues to operate, the site supply will rapidly loose synchronism with the utility company supply. Were the site to remain connected to the external network, then the reclosing of a remote utility company breaker and the unexpected restoration of the external supply could result in severe damage to generating plant. It is vital, therefore, that the site intake breaker is tripped

when a loss of mains occurs. Indeed, where the utility company network incorporates 'auto-re-close' protective equipment, the intake breaker must be opened within the open and reclose time of the remote circuit breaker, which can be less than 0.3 seconds.

The technique used to provide this 'Loss of Mains Protection' will differ according to the circumstances at each site. A discussion of Loss of Mains Protection is given in insert panel 9 of this chapter.

Alternator Negative Phase Sequence Protection Unbalance between the currents drawn in each phase from an alternator leads to the generation of a 'negative phase sequence' component of magnetic flux, which cuts the rotor at twice normal frequency. The resulting currents, when sufficiently large, will cause overheating in the body of the rotor.

Protection against current unbalance due to unbalanced loading on the system or through a phase conductor becoming disconnected may be achieved through the incorporation of 'Negative Phase Sequence Protection' on individual generators.

As far as current unbalance arising from short circuits or earth faults is concerned, alternator protection is usually provided by the operation of Site Supply Protection overcurrent and earth fault relays (see sections 3.8.4 and 3.8.5). Where minimum fault levels are low, however, then the inclusion of 'Phase Unbalance Protection' into the scheme may be necessary. This form of protection is discussed in section 3.8.4.

Generator Synchronising As has been discussed in section 3.7.1, generating sets which use asynchronous alternators require to be connected to an external supply both when they are started and at all times they are generating. The output from such machines is, therefore, unavoidably in synchronism with the external supply.

The same is not true of synchronous alternators, which may be run up to speed and operated in isolation from any other supply. Before a generating set equipped with a synchronous alternator can be connected in parallel with another supply, the following conditions must be met:

(a) The frequency of the output from the set must be the same as the frequency of the supply.
(b) The voltage of the output from the set must equal the voltage on the paralleling bus-bars.
(c) The voltage of the output from the set must be in phase with the bus-bar voltage.

The procedure for meeting these conditions is generally as follows: First, engine governor control is used to bring the set to design operating speed i.e. design synchronous speed for the alternator. Next, alternator excitation is adjusted to match generator output voltage to bus-bar voltage. The difference between the voltage of each phase of the generator output and each phase of the supply is then measured. As the frequency of the generator output at this stage will not be the same as the frequency of the supply, this voltage difference will oscillate from the sum of each pair of voltages to zero. In fact, the frequency of this oscillation will be equal to the difference between generator and supply frequencies.

Engine speed is then further adjusted until the measured voltage difference oscillates slowly, indicating that generator and supply frequencies are nearly matched. Finally, once it has been determined that the two frequencies are sufficiently close, the generator breaker is shut the next time measured voltage difference cycles through zero.

Manual facilities can be provided to execute the above procedure. It is, however, usual for CHP plants to be equipped with programmable logic controllers which are configured to undertake the necessary parameter monitoring and control action.

3.8.4 Site Supply Protection – Generator

The purpose of 'Generator Site Supply Protection' is to prevent damage to the site distribution system and connected loads in the event that electrical faults on individual generators are not cleared by Unit Protection. Some Site Supply Protection measures, however, also act to protect generating plant under certain circumstances, as is discussed in section 3.8.8.

This form of Site Supply Protection is sometimes referred to as 'Load Protection'.

Overcurrent Protection Protection against short circuits or earth faults that are not cleared by individual generator Unit Protection is provided by overcurrent relays. Overcurrent relays and their associated current transformers are located on the output from individual generators, to trip the breaker of the associated generator under fault conditions. Unlike Unit Protection, overcurrent relays do not operate immediately but are designed to have an operating time which is inversely proportional to excess current i.e. the larger the current the faster the operation of the relay.

Overcurrent relays will be used elsewhere on the site distribution system where high voltage supplies are concerned. When this is the case, overcurrent relay settings must be chosen with particular care if the relay nearest to the location of the fault is to operate its associated circuit breaker first, thereby leaving healthy circuits to continue in operation.

Phase Unbalance Protection When a site is operated in parallel with a utility company supply, fault levels will be more than adequate to ensure that overcurrent and earth fault relays operate quickly to clear short circuits and earth faults. As has been discussed in section 3.8.1, however, when a site is islanded, fault levels may be reduced by a factor of 100.

At certain sites this may mean that it is not possible to select overcurrent and earth fault relays with an operating characteristic that will provide reliable Generator Site Supply Protection under all circumstances. Where this is the case, dedicated 'Phase Unbalance Protection' is included into the scheme. This form of protection is provided by 'phase unbalance relays' and associated current transformers which monitor current in each of the 3 phases from individual generators. Should the current in one phase differ from the current in the other two phases by a pre-set value, then the associated generator breaker is tripped.

Where Negative Phase Sequence Protection is not fitted to generators, Phase Unbalance Protection will also protect the site supply should a phase conductor become disconnected at an alternator or its feeder unit.

Under/Over Voltage Protection For large generating plants and high voltage systems, site

9

TECHNIQUES FOR LOSS OF MAINS DETECTION

The requirement for 'Loss of Mains Protection' is discussed in section 3.8.3. In cases where the export of power from local generating plant into the external network is not contemplated, straightforward reverse power relays may be employed to trip the site intake breaker on loss of external supply. These relays must be chosen and set to ensure that they operate before individual generator breakers trip on under-voltage or under-frequency.

Where both the import and export of power is expected, reverse power relays can, obviously, not be utilised. The detection technique employed for this situation will mainly be dependent upon:

a) The magnitude of the change of load seen by the local generating plant on loss of network supply.
b) The speed at which any utility company auto-reclose equipment is set to operate.

Where an increase of load to several times the capacity of the local generating plant is assured, then it may be possible to rely on intake under/over frequency relays to trip the intake breaker before the reconnection of the external supply, even if auto-reclose equipment with a dead time of as little as 1 second is employed on the external network. Should the load on loss of network supply increase to exceed generating plant capacity by only a small margin, then reliance on under/over frequency relays will only be possible where the external network does not incorporate fast acting auto-reclose circuit breakers.

In practice, if the change in load in the event of a loss of external supply is uncertain, then dedicated loss of mains detection equipment will be employed, regardless of remote auto-reclose breaker dead times.

A widely used detection technique uses rate of change of supply frequency as the monitored parameter. Providing the change in load on local generating plant can be relied upon to be significant and relatively high rates of change of frequency do not occur on the supply under normal circumstances, this form of Loss of Main Protection will prove reliable. Another common approach is to monitor generator output voltage for sudden movements in phase displacement. This type of relay typically requires a change of load on the generating plant in excess of 5% of the nominal load at the time of mains failure.

Where load change on loss of mains is likely to be small, a more sophisticated technique, known as 'reactive export error detection' may have to be employed. In simple terms, this type of device overrides normal alternator excitation control to maintain a set reactive current in the circuit interconnecting with the external supply. In the event of mains failure, the monitored reactive current is no longer maintained at the prescribed value and hence loss of mains is detected.

loads are protected against unacceptably high or low output voltage from individual generators using 'under/over voltage relays'. The voltage transformers for these relays are connected at the output from individual generators and trip the associated generator breaker in the event of unacceptable voltage excursions. This form of protection is sometimes referred to as 'Load Under/Over Voltage Protection'.

Adequate discrimination must be achieved between the operation of this protection and 'External Supply Under/Over Voltage Protection'.

Under/Over Frequency Protection In a similar fashion to Voltage Protection, where large generating plants or high voltage systems are concerned, site loads are protected against high or low frequency output from individual generators using 'under/over frequency relays'. Should frequency, as measured by voltage transformers connected at the output from individual generators, deviate from pre-set limits then the associated generator breaker is tripped.

Once again, adequate discrimination must be achieved between the operation of the 'Load Under/Over Frequency Protection' described above and the operation of 'External Supply Under/Over Voltage Protection'.

3.8.5 Site Supply Protection – Intake

The purpose of 'Intake Site Supply Protection' is to prevent damage to the site distribution system and connected loads should electrical faults arise at the intake, on bus-bars and switchgear or on the external supply network. Some Site Supply Protection measures, however, also act to protect generating plant under certain circumstances, as is discussed in section 3.8.8.

The term 'Consumer's Protection' is sometimes used to refer to Intake Site Supply Protection.

Overcurrent and Earth Fault Protection For low voltage systems, Intake Site Supply Protection may simply take the form of Overcurrent Protection, provided by a moulded case circuit breaker or high rupturing capacity (HRC) fuses.

In the case of high voltage systems, protection against short circuits and earth faults at the intake, on paralleling bus-bars and on the associated switchgear is provided by overcurrent and earth fault (OCEF) relays. These relays, which are connected to current transformers located at the intake to the site, have operating times which are inversely proportional to fault current. On the detection of a fault, the OCEF relays cause the site intake breaker to trip.

Phase Unbalance Protection Where external networks use overhead lines, the temporary loss of a phase is not uncommon. Site supplies can be protected against this eventuality through the installation of Phase Unbalance Protection at the intake to the site. The operation of Phase Unbalance Protection is described in section 3.8.4.

Standby Earth Fault Protection Under normal circumstances, a site supply will be provided with an earth reference by the connection of neutral to earth at some point on the utility company network. Where a site is also to be operated in island mode (i.e. disconnected from the external network) an alternative means of neutral earthing will be required. The options available are discussed in section 5.4.2 of chapter 5.

When a site is isolated from the utility com-pany network, the external supply can not, of course, contribute to fault current in the event of an earth fault. Under these circumstances, the earth fault relays and associated current transformers located at the intake to the site will not detect the fault. To protect paralleling bus-bars, generators and switchgear in this eventuality, current transformers and relays are installed at the point of neutral earthing used when the site is islanded. This form of protection is known as 'Standby Earth Fault Protection'.

3.8.6 External Supply Protection – Customer

To allow a site with generating plant to be connected to their network, utility companies will require site operators to install certain protection arrangements. The primary purpose of these arrangements is not to provide protection to the site distribution system or generators but is to protect the external network and other utility company customers connected to it.

Under/Over Voltage and Frequency Protection Where embedded generating plant has sufficient capacity, it may continue to generate and back-feed into the external network, in the event of a loss of utility company supply. Utility companies are, therefore, concerned that, under these conditions, their other local customers are not exposed to unacceptable voltage or frequency excursions due to poor control of the embedded generating plant.

To guard against this possibility, utility companies generally require customers to install 'Under/Over Voltage and Frequency Protection' at the point of electrical intake to the site. The associated voltage and frequency protection relays are set to trip the site intake breaker should either voltage or frequency fall outside the percentage deviations and time durations prescribed by the utility company.

3.8.7 External Supply Protection – Utility Company

For low voltage supplies, the only protection installed by utility companies at the point of supply may be simple Overcurrent Protection, in the form of HRC fuses. Where a site with embedded generation is supplied at high voltage, on the other hand, the utility company will install a number of external supply protective measures.

Overcurrent and Earth Fault Overcurrent and Earth Fault Protection is provided by OCEF relays and current transformers located on the site side of the utility company breaker. In the event of a fault, the utility company breaker is tripped.

This form of protection is provided whether or not the site served has generating plant which operates in parallel with the utility company network.

Reverse Power/Maximum Export Protection 'Reverse Power/Maximum Export Protection' may be provided by 'reverse power relays' which are chosen and set to operate on the reversal of power or at an agreed maximum level of export from the site into the utility company network. As a back-up to reverse power relays, or in some instances as an alternative, 'directional overcurrent relays' may be used. In either case, the relays are arranged to trip the utility company breaker should export into the external network exceed an agreed limit.

Neutral Voltage Displacement Earth Fault Protection Consider a situation when an external supply becomes disconnected at some remote point on the utility company network, generating plant at a site operating in parallel with the external supply continues to generate and that site remains connected to the network. Under these circumstances, should an earth fault on the external supply network occur, the utility company OCEF relays would not detect the fault.

To protect the external network from this eventuality, 'Neutral Voltage Displacement Earth Fault Protection' is installed by the utility company at the point of supply to the site.

3.8.8 Indirect Generator Protection

The prevention of damage to generating sets under circumstances other than those discussed in section 3.8.3, is achieved as a secondary benefit of the operation of Site Supply and External Supply Protection. These other circumstances are considered in the paragraphs which follow.

Unexpected Islanding The operation of Site Supply or External Supply Protection in a number of situations will result in unexpected islanding of a site. If the generating plant in operation at the time of islanding has insuffi-

cient peak capacity to meet total site demand, then 'real' power overloading will occur. Where Under/Over Frequency Protection is installed at the output from individual generators, however, the actual stalling of generators will be avoided through the tripping of generator breakers on under-frequency.

Overload and Load Shedding At sites which require the maintenance of power in the event of an unexpected loss of external supply but which have insufficient generating capacity to meet total site demand, some form of automatic load shedding control will be needed. Load shedding may be undertaken in two ways. Most commonly, the electrical services at a site are designated as 'essential' and 'non-essential' and are served by separate distribution systems. On the loss of the external supply, all non-essential circuits are automatically isolated from the site supply, leaving only essential loads to be fed by the generating plant.

Alternatively, non-essential loads are shed in a predetermined sequence using local contactors. The number of loads shed is determined by the level of under-frequency of the output from the generating plant. The lower the frequency the greater the number of loads shed.

Connection of Stationary Alternators to a Supply Should the previous measures fail to prevent the stalling of generators, then alternators could come to rest with generator circuit breakers closed. The inadvertent re-closure of the site intake breaker under these circumstances would be likely to cause serious damage to the generators (unless the sets are designed for direct-on-line starting). Where Under/Over Voltage Protection is installed at the output from individual generators, however, this eventuality will be avoided as the generator breakers will already be tripped on under-voltage.

Phase Unbalance Protection Where Negative Phase Sequence Protection is not fitted to generators, Phase Unbalance Protection at the output from individual generators will prevent alternator overheat, should either site loading become unbalanced or a phase conductor become disconnected on the site system or external network.

Alternatively, where neither Negative Phase Sequence nor Phase Unbalance Protection is fitted to generators, intake Phase Unbalance Protection will protect alternators in the event of a phase conductor becoming disconnected on the external network.

3.9 Automatic Control and Monitoring

3.9.1 Automatic Control

Control Functions The functions performed by the automatic controls of a CHP system can be summarised as follows:

- To execute the operating strategy for the CHP plant.
- To automate the operation of the individual components of the system.
- To ensure safe operation and automatic shut down of the plant in the event of component failure.

Operating strategy options are discussed in detail in chapter 8. In simple terms, the strategy chosen should operate the CHP plant to achieve optimum financial performance by determining what power and heat to generate from the plant, what power to import and what heat to generate from conventional boiler plant. Once these figures have been determined, the automatic controls modulate the power and heat produced by the plant, including conventional boilers, to provide the required outputs.

The individual components of the CHP plant, engines, alternators, heat recovery units, burners, cooling towers and heat recovery by-pass arrangements are then individually controlled to deliver the chosen power and heat output at optimum efficiency and to reject unwanted heat.

Finally, critical operating conditions on each component of the plant are continuously monitored and processed to detect abnormal operation. On detection of abnormal operation that is likely to damage equipment or present a potential safety hazard, the CHP plant is shut down in a co-ordinated and safe manner and the alternator is disconnected from the site electrical supply system. In the event of CHP plant shut down, the control system will bring additional conventional boilers on-line to ensure that the site heat demand continues to be met. The site electrical demand will automatically be met by the external supply network when the CHP plant has been operating in parallel mode.

The CHP plant control system must, in addition, be fully interlocked with the electrical protection systems to ensure proper co-ordination and safe operation under all circumstances.

Control Systems In the case of small scale CHP, the engine, alternator and heat recovery equipment, plus all necessary automatic controls, are provided as a single 'off the shelf' package by the supplier.

For larger plants, a bespoke control system is required to provide co-ordinated control of the separately supplied components of the CHP installation. The control system is usually supplied and configured by the generating set supplier to meet the overall CHP operating strategy specified by the client.

High quality and highly reliable microprocessor based programmable logic controllers are used, as the security of service demanded of the control system is extremely high. User access to the system is provided by a keypad and display mounted on the face of one of the control panels. For large plants, the control panels themselves are often located in a soundproofed control room with glazing onto the engine room.

A photograph of a typical generator control and synchronising panel is shown in figure 3.9(a).

Control System Failure As mentioned above, modern generator control systems utilise highly reliable microprocessor based programmable logic controllers. The result of a control failure could, however, be catastrophic if the necessary precautions are not taken. The 'watch-dog' circuit on each microprocessor board used in the control system should, therefore, be continuously monitored for fault condition. In the event of a fault on a board or the failure of the power supply to the control system, electro-mechanical controls are provided to take over control of the generators to ensure an orderly and safe shut down of the plant.

3.9.2 Monitoring and Alarm Reporting

Monitoring On older systems, limited remote monitoring facilities are provided by gauges lo-

(Courtesy of Brush Electrical Machines Ltd)

Figure 3.9(a) *Typical Generator Control and Synchronising Panel*

cated on the front of control panels. Many readings, required on a regular basis for operational purposes, have, however, to be taken from gauges mounted on plant itself. The monitoring of older CHP plants can, therefore, be a time consuming task for site staff.

In the latest CHP installations, all pertinent engine, alternator, heat recovery, boiler and heat rejection equipment conditions are sensed, read automatically and temporarily stored by the control and monitoring system. Access to the data is provided by a personal computer (PC) linked to the system. The temporarily stored data is regularly 'downloaded' from the control and monitoring system to the hard disk drive of the PC for later viewing and analysis by plant operators.

In the case of small scale CHP systems, the PC may be located in an office remote from the site, with a normal speech telephone line and autodial modem being used as the communications link. This type of remote monitoring system is often used by suppliers of CHP systems who are contracted to undertake the entire op-

eration and maintenance of the plant.

Alarm Reporting Conventional warning and fault lamps are provided on the control and electrical switchgear panels to advise operators of plant failures. Major faults will also operate an audible alarm. At sites where engine control rooms are not continuously manned, repeater warning and fail lamp panels plus audible alarms will be provided at a point that is always attended.

On modern systems which are equipped with a PC, alarm conditions will also be displayed on the VDU of the computer, logged on the disk drive and usually printed out with the date and time of occurrence. The PC and printer can, of course, be located at any convenient point at the site.

Where operation and maintenance of a small CHP system is provided by a third party with no site based staff, alarms can be reported to a PC located at the company's offices, for immediate acknowledgement and response.

4

ASSESSING THE OPPORTUNITY AND SELECTING A SYSTEM

*Once the potential for CHP at a site has been
confirmed, the detailed work can then start to assess
accurately the opportunity and to make fundamental
decisions on the selection of system components.*

ASSESSING THE OPPORTUNITY AND SELECTING A SYSTEM

4.1 Engine Fuel

Factors to Consider The very first issue that requires consideration is the choice of fuel on which to run the engine of the CHP plant. The use of waste fuels or decomposition gases will, of course, be limited to systems that are to be located adjacent to the requisite installations. In such cases, the opportunity for combined heat and power will be centred around the availability of suitable waste fuel or gases. The decision on engine fuel will, thus, in effect already have been taken.

At most sites, however, a choice will have to be made between one form of fossil fuel or another. The factors that need to be taken into consideration are listed in table 4.1(a). Also given in the table are references to the relevant sections in this book.

The Importance of Natural Gas to Combined Heat and Power As a general rule, when natural gas is available and its cost per unit of energy is comparable with that of the alternative fuel oil supplies, then the convenience and lower maintenance costs associated with gas make it the first choice. Even when plants in excess of 5 MWe are being considered, making heavy fuel oil firing of IC engines and gas turbines an option, the deregulation of the gas supply industries in many countries has made gas prices competitive with heavy fuel oil where large supplies are concerned.

Although natural gas is generally the first choice fuel for CHP installations, its non-availability at a site does not necessarily preclude the development of a cost-effective CHP scheme. In the case of waste fuels and decomposition gases, the increased capital and operating expenditure concerned can be offset against the low or zero cost of the fuel. In addition, in some countries tax and other incentives are available to encourage investment in schemes that do not utilise fossil fuels.

At sites which have no natural gas supply, the use of alternative fossil fuels to fire a CHP plant may prove cost effective under certain circumstances. The general rules of thumb are as follows:

- At sites that utilise light fuel oil for heating and, due to their small size, pay high unit charges for electricity, light fuel oil fired CHP plants can be cost-effective.
- At sites that utilise heavy fuel oil or coal for heating and, due to their large size, pay low unit charges for electricity, only heavy fuel oil fired CHP plants are likely to prove cost effective.
- At sites which have an annual average electrical demand of less than 1,000 kW but utilise heavy fuel oil or coal for heating, the development of a cost effective combined heat and power scheme is unlikely. This is due to the difficulty of finding an IC engine or a gas turbine that can operate satisfactorily on heavy fuel oil in the size range required.

Table 4.1(a) *Choice of Engine Fuel – Factors to be Considered*

Factor	Engine Technology	Section of Book to Reference
❏ Fuel Cost	–	–
❏ Size and type of engine that can be operated on the fuel	IC Engines Gas Turbines Steam Turbines	3.1.3 3.2.3 3.3.3
❏ In the case of gaseous fuels, the pressure at which the gas is available	IC Engines Gas Turbines All	3.1.3 3.2.3 3.6
❏ The capital and maintenance cost implications of operating an engine on the fuel	IC Engines Gas Turbines Steam Turbines	3.1.3 3.2.3 3.3.3
❏ Emissions legislation and regulations	All	5.9.1

Where no gas supply exists, it is always worth discussing the possibility of bringing a supply to a site with the local gas supply company. The supply company may be interested in providing the extension to the supply network at a discount, as an incentive to switch the site to gas firing. The CHP scheme may well then be able to support the remaining capital requirement. Further, in common with electricity utilities, gas supply companies are often interested in joint venture CHP projects and so may be prepared to fund the entire cost of the necessary network extension, along with partial or complete funding of the CHP installation, in return for a share of the cost savings achieved at the site.

Interruptible Gas Supplies plus Run and Standby Fuel Operation At large sites served by gas supply networks that are near to their maximum supply capacity, the option to purchase gas on an interruptible contract can be attractive. Under such contracts, the supplier of gas may elect to cease the supply of gas to a site for a period of days or even weeks at times of extraordinary demand on the network. In return, the customer is able to purchase gas at highly competitive prices. As IC engines (dual fuel design not spark ignition), gas turbines and boiler plant serving steam turbines can all be supplied with the facility to operate on both gas and fuel oil, temporary gas supply interruptions are not a problem, as engines are simply switched over to fuel oil. Sufficient supplies of fuel oil do, of course, have to be maintained in tanks at the site for this eventuality.

The operation of the CHP plant on light fuel oil may, however, prove more expensive than the importation of electricity and the firing of conventional boiler plant. Where this is the case (and where alternative electricity and heat supplies are available) there will be no point in continuing to operate the CHP plant in the event of a gas supply interruption. For this reason, the specification of an engine which can run on either gas or oil will often not be justified, unless the plant is also to serve a standby power function.

Gas Supply Pressure This issue is discussed at length in section 3.6. It is, nevertheless, worth emphasising the importance of utilising as high a supply pressure as possible when a CHP plant

is to use either a dual fuel IC engine or a gas turbine. A fuel gas compression installation for a 5 MWe gas turbine may cost in excess of £100,000 and can consume up to 5% of the power generated by the engine. Investment in new pipework to take a gas supply from a higher pressure point on an external supply network enables compression plant to be scaled down, with the result that capital and operating costs can be significantly reduced. The option of using a high pressure external gas supply should, thus, always be fully investigated with the local gas supply company at an early stage in project development.

4.2 Site Demand Profiles

In chapter 2, average figures for 'major use hours' and 'out of hours' demand in winter and summer are used to test crudely the potential for CHP at a site. Average figures will not, however, reveal the fact that the peaks and troughs in demand for electricity may not coincide with the peaks and troughs in demand for heat. As section 2.2.3 of the chapter emphasised, the convergence of the two demand profiles is vital to the economics of combined heat and power. An accurate assessment of potential financial returns must, therefore, take a close look at how the demand profiles for heat and electricity match up with each other, in addition to considering the demands individually.

This section sets out guidelines on the level of data collection that will be required at sites with various energy use characteristics. In addition, the methods by which the necessary data can be gathered are reviewed.

4.2.1 Constant Demands

In certain circumstances, a CHP plant may be sized to meet the heat demand of a particular process at a site. Where the demand for this heat is almost constant throughout the year and the site demand for electricity is always likely to exceed the power output from the CHP plant, then a detailed knowledge of site demand profiles is not necessary. In such cases, the analysis work required is greatly simplified as adequate accuracy can be achieved using simple average demand figures.

In practice, however, examples of this special

situation are virtually non-existent. The reason is that whilst one process or service at a site may well have a near constant demand for heat, there will be other variable demands for heat that could, in part, be satisfied by the CHP system. So, whilst the sizing of a CHP plant only to meet the constant heat load at a site makes the appraisal work easy, a scheme designed on this basis will almost certainly not achieve the maximum possible return on investment at the site. The issue of plant sizing is considered in detail in section 4.5.

4.2.2 Regular Weekly Demand Profiles

At industrial process sites, loads that vary with outside conditions, such as space heating, may be small in relation to process loads. Where production volumes and patterns are also relatively constant throughout the year, the demand profiles for heat and electricity will follow a regular weekly pattern.

In these circumstances, the determination of a single average weekly profile for heat and for electricity will be all that is needed for the analysis work. Common sense will, of course, have to be used in deciding whether the variation of weekly profile during the year is too great for just one average week to be considered. As a rule of thumb, if site energy consumption on heating fuel for the three winter and three summer months is within 10% and the same is true for electricity and if production volumes and patterns are relatively constant, then the 'regular weekly demand profile' approach can be taken.

4.2.3 Variable Weekly Demand Profiles

At most sites weekly demand patterns are not regular but vary through the year due to changes in outside conditions and alterations to production volumes and work schedules.

In cases where electrical and heat loads can be considered to comprise a fixed element plus an element that can be related approximately to outside air temperature, then an absolute minimum of two weeks' profiles need to be considered. One week's pair of profiles will need to be representative of typical winter demand patterns for electricity and heat, whilst the other week's profiles will be representative of typical summer demand patterns.

As will be seen in section 4.7, however, the summer/winter two week approach will not be adequate in a number of situations. It is, thus, generally recommended that demand profiles for electricity and heat are determined for a minimum of 4 weeks, one week to represent each of the four seasons.

At sites where changes to production volumes and work schedules make it impossible to consider loads as part fixed and part weather related, then it may be necessary to consider more than 4 typical weeks. Again, common sense will need to be used to determine the minimum number of weeks to give adequate accuracy. The objective under these circumstances will be to determine weekly profiles for electricity and heat that are representative of each of the different permutations of loads that exist at the site. For example, if there is a production shut down in summer, then a set of weekly profiles will be needed for summer 'with production' and a second set of profiles will be needed for summer 'without production'.

4.2.4 Tests, Measurement and Data Logging

Regardless of what the demand profiles are thought to be at a site, investigations will be required to confirm and determine accurately the patterns of demand. The required test, measurement and data logging techniques will be well known to those who specialise in energy efficiency. A brief run through is given here for those who may be less familiar.

Heat Demand Where the heat load in question is served by a dedicated boiler plant, then local oil or gas meters can be read at various times during the day for a period of seven days to build up a crude demand profile. Better still, if the meters have been supplied with a 'pulsed output' or if it is possible to attach an optical pick-up to generate a pulse, then a portable data logger can be used to log consumption over as many weeks as are required.

To turn the fuel meter data into actual heat demand, a knowledge of boiler efficiency is required. A portable electronic flue gas analyser will give an instantaneous combustion efficiency reading. This must be coupled with an estimate for boiler shell, idle flue and cycling losses due to purging, to give an annual overall generating efficiency. A figure of 3 to 5% is typical for a 3 boiler installation where 2 boilers are needed at times of peak demand. A word of caution here – many electronic flue gas testers already take off a nominal figure (usually 3%) from the combustion efficiency reading before displaying an answer, to allow for typical boiler losses.

The estimate for generating efficiency can then be applied to the logged fuel consumption data to produce heat demand profiles.

Where fitted, heat meters or steam meters can be read or automatically logged to produce heat demand profiles directly. *Extreme caution* should, however, be exercised as both heat meters and steam meters are notoriously inaccurate when not regularly maintained and recalibrated. Spot checks of accuracy should be undertaken using the techniques outlined above, before credence is given to this type of meter reading.

At some sites heat is imported in the form of steam or hot water generated elsewhere. In the fairly unusual situation where CHP is to be considered at such a site, rather than at the point of heat generation, then metering of the heat supplied will be required. Where the necessary heat or steam metering facilities do not exist, *they must be installed*. As the investment required even for a 100 kWe CHP system will be upwards of £50,000, a few thousand pounds spent on metering equipment to produce reliable heat demand data, will always be well justified.

Where heat and steam meters exist, as mentioned above, they must always be checked for accuracy before use is made of any readings taken. In the case of heat meters, the accuracy of the flow rate reading should be confirmed by disconnecting the pressure tappings on the flow measurement device and taking an independent differential pressure reading preferably using a simple manometer. This can then be applied to the pressure drop versus flow rate characteristic for the measurement device to determine volume and hence mass flow rate.

Where a heat meter uses a turbine or ultrasonic flow measurement device, use should be made of an existing orifice plate or pressure tapped valve in the pipework to check mass flow rate. In the last resort, where measuring stations do not exist and shutting the heat supply down to fit a suitable orifice plate is not possible, a non-invasive Doppler meter can be used to give a rough check of water velocity and hence mass flow.

Finally, the accuracy of the differential temperature measurement used by the heat meter should be verified using a digital thermometer. The thermometer need not be externally calibrated but the same thermometer must be used to take both flow and return water temperature readings, as the temperature differential is the important consideration rather than absolute temperature.

In the case of steam meters, steam condition at the point of supply must first be confirmed. This can be done by using temperature and pressure readings taken at the boiler plant by the supplier of heat, with due allowance being made for heat loss from the pipework up to the point of supply to the site. Pressure drop readings taken on the measurement device of the meter or on a nearby orifice plate can then be combined with known pressure drop characteristics for the chosen device to produce a flow rate reading. Mass flow rate can then be worked out, based on the estimated steam condition.

Where the steam meter uses a turbine flow measurement device and no suitable orifice plate or venturi exists, an estimate of steam rate can be made from condensate return rate. A drain cock will need to be fitted to the condensate line and condensate collected over a given period of time. Caution should, however, be exercised in the use of the data collected as condensate may not be returned from all items of equipment and some water may be lost to steam leaks.

At most sites, different services and processes will require heat in different forms and at different temperatures. Where a load is significant, say 20% or more of total site load, then separate measurements and data logging will be required for that load. The importance of having knowledge of the different heat loads at a site is discussed in section 4.3.

Often, sub-metering facilities for separate loads will not exist. Where this is the case, the installation of the necessary metering equipment to record demand will be justified in most cases. When load is small and investment in sub-metering cannot be justified, then the approximate techniques described earlier in this section can be used to get a rough idea of the demand profile.

Electricity Demand The use of electronic metering equipment by electricity utilities is be-

1

ALTERNATIVE SCHEMES AND ALTERNATIVE FINANCING

Alternative Schemes Large capital projects are attractive both to those who are responsible for energy expenditure and to senior management, due to the kudos that can be generated by a large and successful capital scheme. At the present time, combined heat and power is viewed as being particularly attractive in a number of Western countries. Whilst the interest and support of senior management in energy schemes can only be welcomed, there is a risk that in pressing ahead with CHP many far more cost effective opportunities to cut energy costs are overlooked.

Before a single data logger is deployed to record a demand profile a deep breath should be taken, followed by the execution of a comprehensive review of the current position regarding energy efficiency at the site in question.

At sites where a formal energy efficiency policy and programme is in place, the necessary facts will be readily to hand. Any outstanding schemes with paybacks of less than 3 years should, almost certainly, be funded and implemented before CHP. Schemes with a 3 to 5 year payback will be in direct competition for funds with a CHP scheme. Finally, a knowledge of the likely impact on site demands of energy schemes already in the pipeline but not yet implemented will be vital to the CHP scheme appraisal.

At sites that do not have a formal policy and programme for energy efficiency, the installation of a combined heat and power plant will almost certainly not be the most cost-effective potential energy measure that exists at the site. In fact, at some sites the implementation of all energy efficiency schemes with a 0 to 3 year payback will reduce the demands for heat and electricity to such a degree that CHP is no longer cost-effective.

Alternative Financing With funds for non-core business projects being tight in companies world-wide, people are naturally turning to alternative methods of financing. In the case of energy efficiency, the financing of capital projects by equipment suppliers and 'contract energy management (CEM)' companies is now commonplace.

In essence, the equipment supplier/CEM company raises capital from a third party, provides all the expertise and finance for the capital scheme and takes a share of the cost savings achieved. The various types of arrangement available are discussed in section 6.1.

Herein lies the danger. In companies that have particularly tight constraints on finance there will be a temptation to go straight to an equipment supplier or CEM company to achieve energy savings, before developing a policy and programme for the site. In the case of equipment suppliers they are, quite naturally, only interested in the installation of their equipment and simply will not investigate alternative energy efficiency schemes at the site. A CEM company can, of course, be expected to consider a range of opportunities for the site. Experience shows, however, that the predisposition of most CEM companies is towards large capital schemes and not the 'nitty gritty' of no and low cost measures. This is probably the direct result of the nature of CEM agreements, which mean that the engineering and management resources of the CEM company will achieve greater financial returns when directed towards large capital schemes. Competitive tendering of a site to several CEM companies will make no difference.

There is, therefore, no logical alternative to the development of a policy and programme for energy efficiency at a site (using in-house staff or an independent external consultant) before CHP is seriously considered. It is only by tackling energy efficiency in this structured and logical way that the maximum cost savings can be achieved at a site and the true potential for investment in CHP determined.

coming more and more common. Where this is the case, the utility will be able to provide half hour consumption figures over a complete year either for a small charge or often free of charge. Where such data is available, no further measurement and data logging work is required.

At sites with conventional metering the pulsed output from the meter, or an optical pick-up to generate a pulse, can be logged by a portable logger. The alternative is to place split current transformers over the individual intake cables carrying each phase and connect these to an independent meter and logger. The practical difficulties of gaining access to the intake cables in a safe manner without isolating the supply, make the use of existing metering facilities the preferred option. This appears to be contrary to the advice given in relation to heat and steam meters where independent checking of the accuracy of such equipment is strongly recommended. The reason for this is that the electricity meters used by utility companies are accurate and the technology used means that, by and large, they stay accurate.

There is generally no need to determine the individual load profiles of different processes, as electrical energy transported by cables exists in only one form. The implications of differing supply voltages when a site has two or more separate intakes, are discussed in section 4.4.3.

4.2.5 Estimation of Demands by Calculation

For new facilities, the option to base demand profiles on the findings from site measurements and data logging is obviously not available. When evaluating CHP schemes for new sites, therefore, the magnitude and variation in the demands for electricity and heat have to be derived solely from calculations. For all but the smallest and simplest of facilities it is strongly recommended that use is made of the full dynamic simulation of buildings and plant to assist with the prediction of hourly demands.

For existing sites, though it may be tempting to cut investigative costs by estimating demands purely from calculations, predictions will only be reliable when they are based on the requisite measurement and logging work.

Table 4.3(a)   *Matching CHP Plant with Required Heat Form*

Required Heat Form	CHP Plant Heat Source	Heat Exchange Equipment
Hot water	Engine jacket cooling system (IC engines only)	Shell and tube or plate heat exchanger
Hot water	Lubricating oil cooling system (IC engines only, in practice)	Shell and tube or plate heat exchanger
Steam	Exhaust gases	Exhaust heat exchange unit and steam separator or waste heat boiler
Hot air	Exhaust gases	Waste heat boiler serving a steam heater battery or none (where direct use of exhaust gases is permissible)
Products of combustion (direct fired equipment)	Exhaust gases	None (where it is practicable to use exhaust gases in existing equipment)
Electricity (direct electrically heated equipment)	None	Heat can not be provided by CHP plant unless heating equipment is changed

4.3 Required Form and Temperature for Heat

4.3.1 Form

Heat can, of course, be transported and delivered in many forms. It will be obvious, therefore, that a consideration of heat demand and CHP heat output simply in terms of magnitude will not suffice. The question of whether the CHP plant can produce heat in the form that a particular process requires has to be addressed.

Whilst a knowledge of engineering and heat transfer is required to answer this question in detail, table 4.3(a) gives some guidance.

4.3.2 Temperature

Of equal importance is the temperature at which heat is required. The CHP plant has to be able to produce heat at or above the temperature required for the heating application to be served. Tables 3.1(a), 3.2(a), 3.3(a) and 3.3(b) of chapter 3 give guidance on the temperature at which heat may be recovered from engines of various sizes and types.

4.3.3 Engine Technology

Once the form and temperature of each significant demand for heat at a site has been determined, a table needs to be drawn up to set the data out in a logical form. Tables 4.3(b) and 4.3(c) illustrate how this can be done for two example sites.

As will be seen in section 4.6, the form in which and the temperature at which heat is required has a crucial role to play in the process of engine technology selection.

2

ABSORPTION REFRIGERATION AND CFC's

Absorption versus Electric Vapour Compression Refrigeration To improve the potential performance of a planned combined heat and power scheme, particularly in the summer, it is often worth considering the use of absorption refrigeration in place of electric motor driven vapour compression refrigeration.

For new build projects, when chilled water is to be produced, the additional cost of a lithium bromide chiller is relatively small in comparison to a standard electric machine of the same cooling capacity. As far as maintenance costs are concerned, the difference is again marginal, now that modern designs of lithium bromide chillers have virtually eliminated the old problems of holding the necessary vacuum and avoiding crystallisation.

The advantage of absorption refrigeration where CHP schemes are concerned is, of course, that site demand is switched from electricity to heat. As absorption chillers can be driven either directly by the exhaust gases of the CHP engine or by the heat recovered, the increased use of CHP heat output, especially in the summer when the general demand for heat may be low, can make all the difference to the economics of cogeneration at a site.

The Elimination of CFC's In the case of retro-fit CHP projects, the replacement of existing serviceable electric chillers with absorption machines is unlikely to be justified on the grounds of energy cost savings alone. The signing of the 'Montreal Protocol' and the subsequent phasing out of the manufacture of chlorofluorocarbon (CFC) fluids has, however, given rise to practical difficulties in the continued operation of refrigeration machines that utilise CFC's.

The conventional solution to this problem is to modify the CFC chiller, at some considerable expense, to utilise an alternative non ozone-depleting refrigerant. In some cases, the necessary modifications are so extensive that complete replacement of the machine is more cost effective.

The alternative solution to this problem, where cogeneration is being considered at a site, is to replace the offending vapour compression chillers with absorption machines driven from CHP system heat. Using this route, the CFC difficulty is eliminated whilst the economics of the proposed cogeneration scheme are improved.

Table 4.3(b) *Form and Temperature of Site Heat Demands – Example 1*

Service		Current Heat Demand		Daytime Average Demand (table 2.2(a)) – kW		Potential CHP Heat Source and Demand that could be satisfied – kW					
		Form in which Heat is currently supplied	Temperature at which Heat is currently supplied – °C			Engine Jacket Water		Lubricating Oil		Exhaust Gases	
Ref	Description			Winter	Summer	Winter	Summer	Winter	Summer	Winter	Summer
A	Hot water for bottle washing generated in central calorifier with steam coil served from steam mains	Steam	200	300	300	0	0	0	0	300	300
B	Space heating using steam heater batteries in unit heaters served from steam mains	Steam	200	300	0	0	0	0	0	300	0
C	Milk pasteurisation using milk/LTHW plate heat exchanger, with LTHW generated in central calorifier with steam coil served from steam mains	Steam	200	200	200	0	0	0	0	200	200
D	Milk cooling using milk/chilled water plate heat exchanger, with chilled water generated by electric chillers	Electricity	–	200	200	0	0	0	0	0	0
Totals		–	–	1,000	700	0	0	0	0	800	500

Table 4.3(c) Form and Temperature of Site Heat Demands – Example 2

Service		Current Heat Demand		Daytime Average Demand (table 2.2(a)) – kW		Potential CHP Heat Source and Demand that could be satisfied – kW					
Ref	Description	Form in which Heat is currently supplied	Temperature at which Heat is currently supplied – °C			Engine Jacket Water		Lubricating Oil		Exhaust Gases	
				Winter	Summer	Winter	Summer	Winter	Summer	Winter	Summer
A	Hot water for bottle washing generated in central calorifier with LTHW coil served from LTHW mains	Hot Water	60	300	300	300	300	300	300	300	300
B	Space heating using steam heater batteries in unit heaters served from steam mains	Steam	200	300	0	0	0	0	0	300	0
C	Milk pasteurisation using milk/LTHW plate heat exchangers served from LTHW mains	Hot Water	80	200	200	200	200	0	0	200	200
D	Milk cooling using milk/chilled water plate heat exchanger with chilled water generated by ammonia absorption chillers served from steam mains	Steam	200	200	200	0	0	0	0	200	200
Totals		–	–	1,000	700	500	500	300	300	1,000	700

109

4.4 Transformation of Heat and Power Demands

As was mentioned in section 2.2.5, the demands for heat and electricity at a site are rarely fixed. There may well be opportunities to transfer a demand from electricity to heat. In addition, and probably more importantly, it may be possible to change the form in which heat is supplied or the temperature used.

4.4.1 Transfer of Demand from Electricity to Heat

As electrical power is of more value than heat, switching an existing demand for heat to one for electricity, to try to improve the case for CHP at a site, need never be considered. Where a site has a low heat to power ratio, however, the transfer of demand from electricity to heat can prove crucial to the economics of cogeneration.

In terms of industrial processes, the knowledge needed to assess the technical feasibility of a switch from, for example, electric to steam heating will only be held by experts in that process. In addition, as the process will often be a core activity for the company, a great deal of effort will usually already have gone into optimising the cost efficiency of the operation. In the case of industrial processes, therefore, the op-

portunities to switch from electricity to heat to improve the economics of a CHP system may be limited. They should, nevertheless, be examined.

More likely candidates will be found in the area of centrally generated site services. Table 4.4(a) gives examples of services that can, under certain circumstances, be cost effectively transferred from electricity to heat as part of the implementation of CHP at a site. Insert panel 2 of this chapter discusses one particular opportunity, changing from electric motor driven refrigeration to absorption refrigeration, in detail.

4.4.2 Change of Required Heat Form or Reduction in Required Temperature

In achieving a fit between an engine and the heat demands at a site, it is sometimes necessary to consider the possibility of changing either the temperature at which the heat is supplied or even the form in which the heat is currently provided.

To evaluate the possibilities, it will be necessary to distinguish between the form and temperature at which heat is currently supplied as opposed to the form and temperature at which heat is actually required by the service concerned. To illustrate this point, consider the example site heat demands set out in tables 4.3(b)

Table 4.4(a) *Centrally Generated Site Services that are Potential Candidates for a Transfer from Electric to Heat Operation*

Site Service	Existing Generation Equipment	Modifications Required for Heat Operation
Hot water for washing	Central calorifiers with electric resistance elements	Replacement of resistance elements with hot water or steam coils
Steam for space humidification or sterilisation	Central electrode boilers	Replacement of boilers with steam generators fitted with high temperature hot water or steam coils
Warm air for space heating	Central electric night storage units	Replacement of night storage units with hot water or steam heater batteries
Chilled water for space cooling	Central vapour compression chillers with electric motor drives	Replacement of electric chillers with absorption chillers served by hot water or steam and/or vapour compression chillers with steam turbine drives

and 4.3(c). At the example 1 site, a central steam system and steam mains supply a steam/hot water calorifier to generate water for washing and a steam/low temperature hot water (LTHW) calorifier to generate LTHW for milk pasteurisation. In both these instances, heat is supplied in a different form and at a higher temperature than is required by the service.

At the example 2 site, these same two services are supplied with heat from a LTHW mains. The disadvantage of this approach is that the losses associated with LTHW distribution are greater than those associated with steam distribution. As table 4.3(c) shows, however, the advantage is that the heat required by these services can be generated at a lower temperature and hence may be recovered from the engine jacket water and lubricating oil systems of an IC engine.

Now, in excess of 60% of the total recoverable heat from a typical IC engine is available at relatively low temperatures from the engine jacket water and lubricating oil systems. In many instances, therefore, the case for choosing an IC engine will centre on whether adequate demand for heat at relatively low temperatures exists at a site.

The relevance of this will become apparent when smaller sites are considered. At a site with an annual average electrical demand of less than 500 kW, the use of a gas turbine as the main engine of a CHP plant will, almost certainly, not be economic. For smaller sites, therefore, a fit must be found between site heat demands and the heat recoverable from an IC engine. Otherwise, cogeneration is likely to be a non-starter.

From the above, the importance of examining the way in which heat is supplied and then evaluating the possibilities for changing the form and temperature through system modifications should be clear.

4.4.3 Rationalisation of Electrical Supplies and Distribution Systems
An additional issue that is sometimes of importance concerns the way in which electricity is supplied to and distributed around a site. At large sites that have been developed over a number of years several external supplies may serve a number of separate distribution systems, some at low voltage and some at high voltage. The economics of a CHP scheme for such a site

may require that the CHP plant is able to serve electrical demand throughout the site. Under these circumstances it will be necessary to consider the rationalisation of electrical supplies and distribution systems through, for example, the installation of a high voltage network.

Like all installation work concerned with high voltage equipment, the costs will be significant. Under certain circumstances, however, the combination of improved CHP scheme economics and the lower charge tariff achieved for electricity that will still have to be imported can provide a financial case for the installation of an HV network at a site.

4.5 Engine Plant Sizing

4.5.1 No Export of Heat and Power
Before heat and electricity demand profiles can be used to determine the economics of applying combined heat and power to a site, fundamental decisions have to be taken on the sizing of the CHP plant. This is a 'catch 22' situation, as the optimum decision on sizing cannot be taken until the potential financial returns are known and these cannot be calculated without first knowing what the maximum heat and power outputs of the CHP plant are. The inescapable conclusion, therefore, is that CHP plant selection and sizing is a process of iteration. This is discussed in more detail in section 4.10.

The alternative criteria for sizing the heat engine of a CHP plant are given in table 4.5(a). Also given in the table are typical applications for which each sizing criterion is used. It should be noted that, with the exception of two special cases, site electrical demand considerations dominate the selection process. This is due to the fact, stated regularly throughout this book, that power is of more value than heat. At most sites the achievement of an adequate return on the investment in a combined heat and power scheme will depend on near full use being made of the electrical generating capacity of the CHP plant for more than half of the hours in the year.

Further, at sites where the demand for both heat and electricity varies during the week and from season to season, "criterion C" from table 4.5(a) will apply. This, unfortunately, means that at many sites that have potential for CHP there is no simple criterion which can be used to size the heat engine of the plant, making an

Table 4.5(a) *Alternative Criteria for Engine Sizing – No Export of Heat or Power*

Ref	Criterion	Typical Application
A	Maximum engine power output to equal minimum site electrical demand	A site that is in continuous use but has peaks of electrical demand of relatively short duration
B	Maximum engine power output to equal maximum site electrical demand	A site where the CHP plant is to provide total standby power generation in the event of external supply failure
C	Maximum engine power output to be somewhere between minimum and maximum site electrical demand	Most sites
D	Maximum engine heat output to equal minimum site heat demand at times of day rate electricity charges	A small site where the scale of the CHP project doesn't warrant the provision of alternative heat rejection equipment for the IC engine
E	Maximum engine heat output is equal to maximum heat demand for a particular process at a site	Gas turbine exhaust used directly for process drying

iterative process of size optimisation unavoidable. An engine sized to give maximum power output equal to the annual average electrical demand at the site is, however, a good starting point.

4.5.2 Export of Heat and/or Power

The economics and hence sizing of plant for CHP schemes where the export of heat or power is envisaged will depend, to a significant extent, on the electricity supply industry regulatory arrangements in force in the country concerned. Notwithstanding the economic distortions this may cause, the fundamental criteria are similar to those that apply to CHP schemes where no heat or power is to be exported. Rough guidance is given in table 4.5(b).

4.6 Preliminary System Selection

This section provides guidance on preliminary system component selection. Once initial cost benefit calculations have been undertaken it may be necessary to change some of these decisions. This, however, is simply part of the iterative process of scheme optimisation.

4.6.1 Engine Technology

The overriding factors when it comes to choosing engine technology are engine power output and the temperature at which heat is required. The flow chart given in figure 2.6(a) of chapter 2 is provided to assist the reader with engine technology selection.

4.6.2 CHP Plant Package or Individual Component Selection?

For CHP plants with an electrical output of less than 1 MW there are manufacturers who provide packaged CHP plants (engine, alternator, heat recovery equipment, automatic controls and acoustic enclosure) skid mounted ready for shipping. In terms of having a single party responsible for the design, component matching and supply of equipment the advantages of the packaged approach are significant. For this reason, where such packages are available they should always be considered in preference to the bespoke assembly of a CHP plant.

For larger plants, where individual component selection is unavoidable, guidance on the selection of the ancillary components is given in the sections which follow.

4.6.3 Heat Recovery Equipment

For engine jacket water and lubricating oil cooling systems, suitable heat exchangers will usually be provided as part of the generating set package. Sizing and selection is not, therefore, a concern of the purchaser.

An exhaust gas waste heat boiler will, how-

Table 4.5(b)   *Alternative Criteria for Engine Sizing – Export of Heat and/or Power*

Ref	Criterion	Typical Application
A	Maximum engine plant power output to equal minimum total customer electrical demand	The total electrical demand profile of customers served is fairly constant but has peaks of relatively short duration. Power shortfalls are met by importing electricity from another supplier
B	Maximum engine plant power output to equal (with adequate equipment redundancy) maximum total customer electrical demand	The CHP operator provides the sole electrical supply to its customers
C	Maximum engine plant power output to be somewhere between minimum and maximum total customer electrical demand	Most situations where additional electricity may be imported from another supplier and is guaranteed to be available

ever, need to be sourced. For engines above 1 MW in size, both fire tube and water tube designs of boiler may be considered. Where steam is to be generated, a steam space integral to the boiler shell may be appropriate for smaller plants. For large plants, a remote steam separator will be required.

4.6.4 Boost Firing

Where heat demand is likely to exceed recoverable heat from the CHP plant for a significant number of hours per annum, it may prove cost effective to incorporate boost firing into the scheme. The burner equipment required, forced draught fan arrangements etc. will be determined by the waste heat boiler manufacturer to meet the demands of the site.

In the case of IC engines, boost firing need only be considered where multi-megawatt installations are concerned.

4.6.5 Gas Compression Plant

When a gaseous fuel is to be burned, gas compression plant will be required for gas turbines, dual fuel IC engines and some turbo-charged spark ignition IC engines. The major decision to be taken is the choice of compressor technology – screw or reciprocating? The pros and cons of each type of machine have been discussed in section 3.6 of chapter 3.

The conclusion that may be drawn is, whilst either type of compressor can be used to serve the engine of a CHP plant, screw machines have a number of operational advantages. In excess of 5% of the power output from a CHP plant may be required for fuel gas compression, where gas turbines or dual fuel IC engines are concerned. The annual operating efficiency achieved by the gas compression plant can, therefore, have a significant impact on the overall economics of a cogeneration scheme. The appropriate weight should, thus, be placed on operating efficiency, *at the loadings expected*, when the decision between screw and reciprocating compressors is made.

The suppliers of compressors should be asked to return information and prices for a complete packaged gas supply plant, skid mounted ready for connection to the site gas supply, engine fuel intake and electrical supply. In this way, responsibility for the selection and inclusion of all the necessary gear, including safety equipment, will be held by the compressor supplier. In addition, it will be the supplier's responsibility to ensure that any particular requirements of the local gas supply utility are catered for.

Finally, discussions will need to be held with the gas supply utility to confirm that the necessary volumes of natural gas will be available and to determine the point on the local network from which the supply is to be taken. Wherever practical, the external supply should be taken at a pressure that is as close as possible to the pressure required by the engine of the CHP plant to minimise the recompression that has to be undertaken on site. Where a medium pressure external gas supply can be made available, the associated reduction in capital and operating costs for the gas compression plant can be substantial.

4.6.6 Alternators

For virtually all CHP installations the alternator will be supplied as an integral part of a manufacturer's standard generating set package which includes engine, alternator, controls and protection equipment. In exceptional circumstances specially constructed alternators may be needed where unusual fault level constraints exist. This is an area where specialist expertise is required.

Regardless of whether a standard generating set or a bespoke alternator/engine combination is used, the purchaser will not usually be involved in the detailed process of matching an alternator to an engine. Consideration will, nevertheless, need to be given by the purchaser to the issues raised in section 3.7 of chapter 3 to enable the broad selection parameters to be decided. The two fundamental decisions to be taken are generating voltage and whether the alternator is to be synchronous or asynchronous.

As has been discussed in section 3.7, synchronous alternators have operational advantages over asynchronous machines in terms of local voltage and power factor control. On the other hand, the need for synchronisation equipment and additional protection measures along with the greater complexity of the machines themselves, make synchronous alternators more costly to install.

As a simple guide, for small CHP installations, where generator power output is modest in relation to site demand, asynchronous alternators are likely to prove the optimum choice. For larger installations and where more sophisticated CHP plant control strategies are to be employed, the operational flexibility of synchronous alternators will be needed. Finally, if the CHP plant is to continue to operate in the event of a loss of the external supply, then synchronous alternators have to be used (unless other standby generating plant is installed at the site).

In terms of generating voltage, high voltage generation is more efficient but requires greater capital investment. For CHP plant with an electrical output below 1 MW, low voltage generation is usually the right choice. Where electrical output is to exceed 1 MW and a suitable point on an existing high voltage network can be found, then generating at high voltage will probably be the optimum decision.

For plants with an output of around 1 MWe, the capital cost advantage of generating at low voltage and stepping up to high voltage using a delta/star transformer may outweigh the operating cost advantages of high voltage generation.

4.6.7 Electrical Switchgear and Protection

Selection of the necessary electrical switchgear and protection can be a complex and involved task requiring specialist expertise, particularly where high voltage installations are concerned.

For small CHP plant around the 100 kWe size, preliminary selection advice can be sought from CHP package manufacturers. For larger installations, specialist expertise must be sought at an early stage in scheme development to ensure that potential problems are identified early and a good idea of the necessary capital expenditure is determined. A lack of care in this area at an early stage in scheme development can lead to some nasty surprises later on. So beware!

4.6.8 Automatic Control and Monitoring

It is usual for generating set manufacturers to provide the automatic control and monitoring facilities for the entire CHP installation. Their advice should, therefore, be sought regarding the preliminary selection of control and monitoring gear.

4.6.9 Multiple Engine Installations

At sites where demand varies significantly, it might be thought that the case for a twin engined CHP plant could be made – two engines to operate during 'major use hours' and one to operate 'out of hours'. As full load efficiency tends to increase with engine size, however, the difference between the annual generating efficiency achieved by a twin engined installation and an installation using a single engine of twice the size will often be small. This is particularly true for IC engines as they have relatively good part load performance characteristics.

In practice, therefore, the modest annual performance advantages achieved by twin engine schemes are usually not sufficient to outweigh the higher capital and maintenance costs that will generally be associated with two engines as opposed to one engine of twice the size. Reference to the nomographs given in figures 2.6(b) and 2.6(c) of chapter 2 will reveal that the capital and maintenance penalties are particularly severe for CHP plants in the 20 to 100 kWe size range where IC engines are concerned and the 500 to 3,000 kWe size range in the case of gas turbines.

For these reasons, the use of two engines for a cogeneration installation in place of one large engine is unlikely to be justified on economic grounds. Should a CHP plant be required to provide secure supplies of either electricity or heat then, of course, a multiple engine installation, with redundancy of equipment, will be needed.

Where two or more engines are to be installed, individual heat exchangers and waste heat recovery boilers will usually be required for each engine. A common gas compression plant can, however, be used to serve a number of engines. In terms of control, automatic load sharing will need to be incorporated into the overall plant control strategy.

4.7 Prediction of Energy Savings/Revenue

4.7.1 The Approach

Bin Method To predict the energy cost savings/revenue that will be achieved by a CHP plant at any point in time, 3 separate sets of variables must be known. These are:

- The demand for electricity and for heat at that point in time and the control strategy to be applied to the CHP plant.
- The performance of the CHP plant at the outside air conditions and at the loading of the plant at that point in time.
- The fuel tariffs that apply at that point in time.

For installations that are to utilise a sophisticated CHP plant control strategy, a further complicating factor arises in that the control strategy may vary depending upon the values of all 3 sets of variables.

Fortunately, the performance of CHP plant can be considered to change instantaneously from one set of operating conditions to another. For this reason, a set of progressing, time-series

calculations is not required for the evaluation of plant performance. Point-by-point, instantaneous calculations are perfectly adequate.

The approach, therefore, is to identify periods of time in a typical year when all 3 sets of variables can be considered to be constant. Instantaneous calculations are then undertaken for each of these time periods and the results multiplied by the length of time in each period to give total figures for the year. This is commonly known as the 'bin' calculation approach.

Unfortunately, a CHP system comprises a number of interconnected and hence interdependent items of relatively complex plant. For example, the electrical load on the alternator determines the power output from the engine, the engine power output determines the engine fuel consumed and the quantity and temperature of the heat available for recovery and so on. The result of this interdependence is that the evaluation of CHP plant performance simply for a single point in time, demands the execution of a large number of interrelated calculations. The key, therefore, is to minimise the number of time periods or bins for which instantaneous calculations are to be executed, whilst maintaining adequate prediction accuracy.

The sections which follow give guidance on the number of time periods that need to be considered for various combinations of circumstances. The following three rules can, however, be applied universally.

a) Where IC engines or steam turbines are to be used, the variation of engine performance with outside air conditions is so small that it can be ignored. Figures for engine performance for a single representative outside air temperature and pressure can, therefore, be used.

b) At sites where the demand for electricity and the demand for heat vary by less than 10% throughout the year and the demand patterns are roughly co-incident, a single average figure for each demand can be used for the whole year. Under these circumstances, fuel tariff structures will determine the choice of bins.

c) At sites where the tariffs for all relevant fuels are virtually constant throughout the year, a single average figure for each fuel cost can be used. Under these circumstances, site demand profiles will determine the choice of bins.

Load Duration Curve Method It is appropriate to mention here an alternative approach to the sizing of CHP plant and the prediction of energy savings/revenues. The approach makes use of load duration curves for heat and/or electricity. Such a curve is illustrated in figure 4.7(a).

The approach is, however, limited to applications where the CHP plant is always to be run at full load or else shut down. In addition, load duration curves give no indication as to the coincident nature of heat and electricity demands, a factor that is crucial to the economics of CHP. Finally, the actual construction of load duration curves is a complex and lengthy task in itself.

For these reasons, the load duration curve method is not discussed in this book and is not recommended.

4.7.2 The Number of Bins Required

Small Scale CHP As discussed in section 3.1.4, for small CHP systems of less than 100 kWe output the provision of alternative heat reject facilities is unlikely to be cost effective. So, for this type of installation, the CHP plant can only run when site heat demand exceeds the required

rate of heat rejection from the engine (excluding unrecovered exhaust heat). The only exception to this would be an installation which had been provided with the necessary controls to vary power output to match heat recovery to site heat demand. For small CHP schemes, however, the added complexity would be unlikely to be justified by the additional savings achieved.

For a scheme with no alternative heat rejection facilities, the sizing of CHP plant is dominated by the requirement for site heat demand to exceed engine heat output for a sufficient proportion of the year to achieve an economic number of engine running hours per annum. The consequence of sizing plant on this basis is that, for most sites, the chosen engine will have an electrical power output which is small in relation to site electrical demand.

Finally, the comparatively high fuel and maintenance costs associated with small CHP schemes make the operation of the plant at times of low, night-time electricity unit charges marginal at best.

From the above it will be concluded that this application falls under "reference D" of table 4.5(a), where the criterion for engine sizing is "Maximum engine heat output to equal minimum site heat demand at times of day rate electricity charges". This type of cogeneration system is designed to operate at full power during the day with all electricity and heat being usefully used at the site. At night the system is shut down.

For small scale schemes, therefore, the performance of the CHP plant can be calculated for a single point in time, using full load performance figures for each item of equipment. The number of bins required will, thus, be determined purely by the number of different fuel tariff charges that apply through the year. The affect of varying outside air conditions on engine performance can be ignored as small scale CHP plants use only IC engines.

Typically, calculations for a total of 4 bins will need to be executed (one set of calculations for each quarter of the year) to account for any seasonal variations in fuel tariffs.

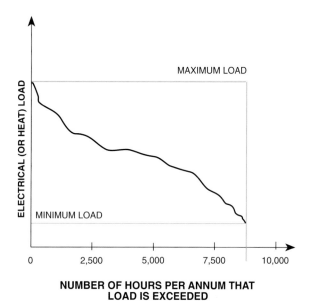

Figure 4.7(a) *Load Duration curve*

CHP Plant Electrical Output Sized to Equal Minimum Site Electrical Demand The engine of a CHP plant sized on this basis will, of course, operate continuously at full load, provided that site heat demand is always sufficient or heat dump equipment is installed.

When an IC engine or steam turbine is to be used the performance of the engine can be calculated for a single point in time, using full load performance figures at a representative outside air temperature and pressure. The number of bins required to evaluate such a scheme will, thus, be determined by the variation in heat demand profiles and tariff charges.

For gas turbines, however, a single calculation for full load performance will not suffice as maximum output varies markedly with ambient conditions. Where a gas turbine is under consideration, it is necessary to calculate performance for each chosen time period using a representative outside air temperature and pressure for that period. These calculations will show that, at full load, power output and hence electricity generated is not constant but changes

Table 4.7(a) *Guidelines for the Selection of Bins for the Prediction of CHP Scheme Energy Savings/Revenues*

| Ref | Demand Characteristics for each month | Fuel Tariff Characteristics for each month | Bins | |
			No.	Time Periods[1]
A	One demand constant, the other follows regular weekday and weekend day profiles	Heating fuel charges constant, day/night unit charges for electricity	24–48	Major use hours and out of hours periods to represent each month of the year (more periods if day/night charges do not coincide with major use hours/out of hours time split)
B	One demand constant, the other follows regular weekday and weekend day profiles	STOD[2] charges for one or both fuels	>24	As 'A' but with extra periods as necessary to accommodate the STOD charge bands
C	Both demands follow regular weekday and weekend day profiles and the demand patterns are co-incident[3]	Heating fuel charges constant, day/night unit charges for electricity	24–48	As 'A'
D	Both demands follow regular weekday and weekend day profiles and the demand patterns are co-incident	STOD charges for one or both fuels	>24	As 'B'
E	Both demands follow regular weekday and weekend day profiles but the demand patterns are not co-incident	Constant or STOD fuel charges	96	4 major use hours and 4 out of hours periods to represent each month of the year (chosen to match STOD charge bands where applicable)
F	Demands do not follow regular weekly profiles	Constant or STOD fuel charges	>96	The minimum number of periods that can be selected to represent each major combination of demands and fuel tariffs

Notes
[1] Major use hours and out of hours periods are defined in table 2.2(a) of chapter 2.
[2] STOD – seasonal time of day tariff, often used for supplies of electricity.
[3] Co-incident demand patterns mean that the troughs and peaks in each demand occur at approximately the same times.

significantly from winter to summer and from daytime to night-time.

Whichever type of engine is used, a minimum of 8 bins will, typically, need to be considered. These will comprise 4 periods to represent average 'major use hours' heat demand and fuel tariffs for each quarter of the year and 4 periods to represent average 'out of hours' heat demand and fuel tariffs. Any variations in electricity unit charges through the day which do not roughly coincide with the 'major use hours/out of hours' time split, may dictate that additional time periods have to be considered.

All Other Situations In all other situations there are no generalised simplifications that can be employed. A degree of judgement, therefore, needs to be exercised in the determination of the number of bins for which calculations need to be undertaken to achieve adequate energy cost savings/revenue prediction accuracy. The factors that need to be considered and rough guidelines are given in table 4.7(a).

4.7.3 Input Data Required

Heat and Electricity Demands The required data on heat and electricity demands has been discussed at length in section 4.2 of this chapter.

Engine Performance Advice on the choice of engine technology and the determination of engine size has been given in section 4.6. With this preliminary selection made, a manufacturer must then be approached to provide the necessary engine performance data.

As has been discussed in section 4.6.2, for CHP plants with an electrical output of 1 MW or less packaged CHP plants are available. The manufacturers of such packages can be approached for data on the performance of the entire CHP plant. Table 4.7(b) details the information that will be required.

Where the engine is to exceed 1 MW power output, it may not be possible to source a packaged CHP plant. In this case, a manufacturer that packages generating sets should be approached for data on combined engine and alternator performance. Table 4.7(c) details the information that will be required.

Where the engine of the CHP plant is to be a

gas turbine, the performance data detailed in either table 4.7(b) or more likely table 4.7(c) will be required for a range of outside air temperatures, to account for the variation of engine performance with outside air. Data for (say) four representative outside air temperatures will usually prove adequate.

Heat Exchanger Performance When data for a packaged CHP plant has been obtained, then heat exchanger performance will already have been accounted for by the package manufacturer in the calculation of heat recovery rates.

For larger plants, where it has not been possible to source a packaged plant, the performance of suitable heat exchangers will need to be evaluated. In the case of jacket water and lubricating oil, the generating set manufacturer will have to be approached to make sure that the transfer of heat from the engine cooling water and oil circuits to the proposed heating circuits is a practical proposition in terms of temperature difference. Provided a suitable shell and tube or plate heat exchanger can be sourced, it may be assumed that 100% of the heat available from the engine jacket water and lubricating oil systems is recoverable.

The required exhaust gas heat exchanger/waste heat boiler will generally be a bespoke item and so will be designed around the exhaust gas characteristics of the engine concerned. The chosen heat exchanger manufacturer will, therefore, have to be furnished with the relevant data. Table 4.7(d) details the information that will need to be given to manufacturers and the information to be returned by them.

Boost Fired Waste Heat Boiler Performance Table 4.7(e) details the information that will need to be given to manufacturers and the information to be returned. In the case of IC engines, the lower oxygen content of the exhaust gases will usually mean that a forced draught fan system has to be incorporated into the waste heat boiler, to provide adequate oxygen for combustion.

Gas Compressor Performance Where data for a packaged CHP plant has been obtained using table 4.7(b) then gas compressor performance, where relevant, will be provided by the package manufacturer.

In the case of larger plants, where it has not been possible to source a packaged plant, the performance of a suitable gas compressor will need to be evaluated, where one is required. Advice on the choice of gas compressor technology has been given in section 3.6 of chapter 3. Table 4.7(f) details the information that will need to be given to compressor manufacturers and the information to be returned.

Conventional Boiler Plant Performance In cases where existing or new conventional boilers are to be used to supply heat demand in excess of CHP plant capacity, then an estimate of overall boiler plant heat generating efficiency will have to be made. Such an estimate will need to take account of the following losses that are associated with conventional boiler plant:

a) *Shell losses* i.e. the radiant and convective heat losses from the outer surfaces of each boiler.

b) *Idle flue losses* i.e. the convective heat losses due to air passing through each boiler when it is not firing, as a result of natural convection or the operation of an induced draught fan.

c) *Cycling losses* i.e. the convective heat losses due to air passing through each boiler before and after each operation of

Table 4.7(b)   *Performance Data required for Packaged CHP Plants*

Data	Percentage of Generating Set Electrical Output			
	100%	75%	50%	25%
Alternator Electrical Output – kW		–	–	–
Engine Fuel Consumption Rate[1] – kW				
Heat Rejected from Engine Jacket Cooling System[2] – kW				
Heat Rejected from Lubricating Oil Cooling System[3] – kW				
Heat Recovered from Exhaust Gases[4] – kW				
Compressor Electrical Demand[5] – kW				

Notes

[1] At lower heat value of fuel.

[2] The off-engine jacket water temperature must be sufficiently high for the water circuit which is to be heated. A 5°C temperature differential is adequate.

[3] The off-engine lubricating oil temperature must be sufficiently high for the water circuit which is to be heated. A 5°C temperature differential is adequate.

[4] The return temperature, mass flow rate and details of the fluid to be heated has to be given to the CHP plant manufacturer for this data to be ascertained.

[5] Where a fuel gas compressor is required, account must be taken of the electricity it absorbs. Care should be taken to ensure that the CHP plant manufacturer has not already deducted compressor power from the stated electrical output of the CHP plant.

Table 4.7(c) *Performance Data required for Generating Sets*

Data	Percentage of Generating Set Electrical Output			
	100%	*75%*	*50%*	*25%*
Alternator Electrical Output – kW				
Engine Fuel Consumption Rate[1] – kW				
Engine Jacket Cooling Water System:				
Heat rejected – kW				
Off-engine cooling water temperature – °C				
Cooling water mass flow rate – kg/s				
Cooling water specific heat capacity – kJ/kgK		–	–	–
Lubricating Oil Cooling System:				
Heat rejected – kW				
Off-engine oil temperature – °C				
Oil mass flow rate – kg/s				
Oil specific heat capacity – kJ/kgK		–	–	–
Exhaust Gases:				
Temperature at exit from engine – °C				
Mass flow rate – kg/s				
Specific heat capacity – kJ/kgK				

Notes

[1] At lower heat value of fuel.

the burner, as a result of the burner purge sequences.

d) *Combustion losses* i.e. the heat lost in exhaust gases when a boiler is firing.

Assistance from the boiler manufacturer should be sought when estimating these figures for annual average boiler performance. Care should also be taken to allow for the effect that the reduced load on the plant (as a result of the CHP installation) will have on these losses.

Fuel Tariffs Fixed, maximum demand and energy unit charges will need to be known for the following:

Purchased electricity
Exported electricity (where export is contemplated)
Engine fuel
Waste heat boiler boost fire fuel
Conventional boiler fuel
Exported heat (where export is contemplated)

It is likely that current purchase tariffs will change once the CHP plant is operational. For example, purchased electricity may cost more as the volume to be purchased will reduce and natural gas may cost less as the volume to be purchased will increase. As the relative pricing of electricity and heating/engine fuel is the fundamental factor which determines the cost effectiveness or otherwise of a CHP installation, budget quotations will be required from utility companies at an early stage in scheme evaluation.

4.7.4 Bin Calculation Procedure

With all the necessary data to hand, then a prediction of instantaneous energy savings/revenue can be made for each required bin. For ease of use, the data provided by manufacturers should be plotted on graph paper and a smooth curve drawn through the points by hand. The necessary calculation steps are as follows:

Table 4.7(d) *Performance Data required for Exhaust Gas Heat Exchangers/Waste Heat Boilers* P

Data	Percentage of Generating Set Electrical Output			
	100%	*75%*	*50%*	*25%*
Exhaust Gases[1]:				
Temperature at exit from engine – °C				
Mass flow rate – kg/s				
Specific heat capacity – kJ/kgK				
Heat recovered from Exhaust Gases[2] – kW				

Notes

[1] Information to be provided *to* the heat exchanger manufacturer.

[2] The following fixed items of data will also have to be provided to the heat exchanger manufacturer:
  a) Nominal return temperature, mass flow rate and details of the fluid to be heated.
  b) Minimum exhaust gas temperature that is acceptable under normal operating conditions (see section 3.4 of the book).
  c) Heat exchange approach temperature differential. Advice should be sought from the manufacturer on the approach temperature differential for this particular set of parameters that is likely to prove cost effective.

Table 4.7(e) *Performance Data required for Waste Heat Boilers with Boost Firing*

Data	Percentage of Generating Set Electrical Output			
	100%	*75%*	*50%*	*25%*
Exhaust Gases[1]:				
Temperature at exit from engine – °C				
Mass flow rate – kg/s				
Specific heat capacity – kJ/kgK				
Oxygen – %				
Heat Recovered from Boost Fired Exhaust Gases at the following Boost Fire Rates[2]:				
Zero boost fire – kW				
25% of maximum – kW				
50% of maximum – kW				
75% of maximum – kW				
Maximum boost fire – kW				

Notes

[1] Information to be provided *to* the waste heat boiler manufacturer.

[2] The following fixed items of data will also have to be provided to the heat exchanger manufacturer:
   a) Nominal return temperature, mass flow rate and details of the fluid to be heated.
   b) Minimum exhaust gas temperature that is acceptable under normal operating conditions (see section 3.4 of the book).
   c) Heat exchange approach differential. Advice should be sought from the manufacturer on the approach temperature differential for this particular set of parameters that is likely to prove cost effective.
   d) Maximum boost fire input fuel rate. Advice should be sought from the waste heat boiler manufacturer on the boost fuel input rate necessary to achieve a waste heat boiler output equal to the maximum heat demand at the site. Engine exhaust gas mass flow rate or oxygen content may prove to be the limiting factor.

Table 4.7(f) *Performance Data required for Gas Compressors*

Data	Percentage of Generating Set Electrical Output			
	100%	*75%*	*50%*	*25%*
Engine Fuel Consumption Rate[1] – kW				
Compressor Electrical Demand[2] – kW				

Notes
[1] Information to be provided *to* the compressor manufacturer, at the lower heat value of the fuel.
[2] The following fixed items of data will also have to be provided to the compressor manufacturer:
   a) Gas type, lower heat value and details of composition.
   b) Gas pressure at intake to compressor.
   c) Gas pressure required at intake to engine.

Step 1 For plants that utilise a gas turbine, the maximum electrical power output for the generating set should first be calculated using the chosen representative outside air temperature for the bin.

Step 2 The average heat or electricity demand for the bin should then be used to set engine output, up to maximum output, dependent upon whether a heat or power led control strategy is to be employed. Control strategies are discussed in detail in chapter 8.

Step 3 With engine output set then, for a packaged CHP plant, engine fuel consumption rate, heat recovery rate and compressor electrical demand can be determined directly from the information provided by the manufacturer. For non-packaged plant the performance of the individual components of the system must be evaluated in turn using the information provided by each manufacturer, starting with engine fuel consumption rate and exhaust gas data at the set engine output.

Step 4 With the performance of the CHP plant evaluated, then the necessary rates of fuel purchase and export can be calculated.

Step 5 Finally, the energy unit charges that apply for the bin in question can be used to determine the rate of energy cost savings/revenue generation that is achieved for that particular period of time.

4.7.5 Annual Predictions

Energy Units The calculations detailed in section 4.7.4 will need to be executed for each time period or bin that has been chosen for evaluation. The results from these calculations are then multiplied by the number of hours in the relevant bin and the savings/revenue totals summated for the year.

Maximum Demand Charges Many electricity supply contracts include a maximum demand charge component, which is usually based on the maximum number of energy units supplied in each half hour of a given charge period. When applicable, the impact of the CHP installation on this component of electricity charges will need to be evaluated.

Where a large number of relatively short time periods have been used (or where electricity demand is constant), the maximum demand for purchased electricity for each charge period may be determined directly from the results of the bin calculations. Where a limited number of long time periods have been chosen, however, the figures calculated will represent average electricity purchase requirements rather than maximum demands and a separate evaluation of maximum demand for each charge period will need to be undertaken.

4.7.6 The Effect of Plant Availability on Annual Predictions
No matter how reliable a generating set is, there

will be periods of time when the machine is not in service due to routine inspection and maintenance work. This down time, when the CHP plant is not achieving savings/revenue, must be fully allowed for in the annual predictions.

Energy Unit Cost Savings/Revenue It is recommended that the guaranteed availability figures provided by generating set manufacturers are applied to the energy unit predictions. Naturally, these figures are relatively conservative and may well be exceeded in practice. The maintenance contracts under which the availability guarantees are given, however, often provide for payments by the purchaser to the manufacturer when availability for the year is exceeded, in addition to payments in the other direction when the converse is true. Hence the rationale behind using the guaranteed availability figures.

In using availability figures provided by manufacturers, care must be taken to ensure that the figures include planned as well as unplanned downtime. Reference should be made to insert panel 2 of chapter 3 for a full discussion of this issue.

It should be assumed that outages cannot always be arranged at favourable times as far as purchased fuel charges are concerned. Hence, the annual figure for energy unit cost savings/revenue should simply be multiplied by the average availability figure determined for the chosen generating set to give a corrected prediction for the year.

Maximum Demand Charge Savings In the prediction of savings on maximum demand charges, particular caution is warranted. The CHP plant has only to remain out of service for a continuous period of less than one hour each month at times of peak demand, for no maximum demand savings to be achieved. As the availability achieved even by a high quality IC engine driven CHP plant is likely to be less than 95%, this leaves in excess of 400 hours per annum when the plant will not be in operation.

As a general guide, for small scale CHP plants no maximum demand savings should be predicted. For larger plants, advice should be sought from generating set manufacturers.

As a final point, due to the difficulty of achieving maximum demand savings in practice

using a CHP installation, it may be worth considering a change of electricity supply contract terms. Seasonal time of day contracts usually have little or no maximum demand component and may, therefore, prove advantageous to the operator of a CHP plant.

Fixed Charges Finally, the fixed charge associated with each fuel for each charge period must be summated to give a total for the year for the situation with CHP.

In determining these charges it should be remembered that, for virtually all CHP schemes, alternative arrangements for the supply of the entire site heat and electricity demand will have to be retained along with the associated standing charges for those supplies. Indeed, these standing charges may well increase with the reduced consumption of energy and hence revenue for the utility company concerned.

4.8 Estimation of Operating Costs

4.8.1 Energy
Justification for investment in CHP plant comes from the achievement of energy cost savings and, in some cases, the generation of revenue from the sale of energy. The evaluation of energy operating costs has been considered in detail in section 4.7.

4.8.2 Maintenance
The maintenance of engines is discussed at some length in section 2.4.3, chapter 2 and in chapter 8. It is, however, worth re-emphasising here that engine maintenance is a specialised activity that has to be undertaken by trained and fully equipped personnel. A high quality, well maintained IC engine will have a service life in excess of 10 years. A poorly maintained machine might have to be completely stripped down and re-built after only a few years. In addition, availability guarantees and warranties for engines will always be linked to the execution of maintenance by engine manufacturers or nominated service companies.

For all these reasons, allowance must be made for engine maintenance to be undertaken by others.

To give the maximum reliability to the maintenance cost estimate, guaranteed figures for all-inclusive parts and labour maintenance contracts for a 10 year period should be obtained directly from generating set manufacturers. Such contracts will, as mentioned above, provide availability guarantees and may also include the loan of components, such as a gas turbine rotor, in the event of prolonged downtime.

Charges for maintenance are often quoted in terms of cost per kWh of electricity generated. As maintenance costs are, however, more closely related to actual engine running hours rather than electricity generated, where a plant is not operated continuously higher cost per kWh charges are quoted. For larger engines, particularly gas and steam turbines, maintenance costs will often be quoted in terms of cost per hour operated.

Where a cogeneration system is not to be procured as a single package, the maintenance requirements of the ancillary system components must not be overlooked. These requirements include: the cleaning of heat exchangers and waste heat boilers, the maintenance and resetting of boost burners and controls, the maintenance of heat rejection equipment, the maintenance of ventilation fans, the maintenance of fuel conditioning equipment and the cleaning/replacement of filters.

Great care must be taken to ensure that a quoted all-inclusive agreement for maintenance covers all the items you believe it does. The questions which need to be asked are:

- Does the price for a packaged CHP plant cover the maintenance of the entire plant supplied, just the generating set or the engine alone?
- Does the price for a generating set cover the maintenance of the engine, alternator, controls, switchgear and protection or just the engine?
- What routine maintenance and inspection work has to be undertaken by the customer's own site staff and hence falls outside the scope of the agreement?
- Does the agreement truly include all parts and labour or are there restrictions and exceptions?

4.8.3 Operation
Modern CHP installations are designed to be fully automatic and do not require permanent attendance by site staff. As mentioned in the last section, however, there may be some maintenance and inspection duties which have to be undertaken by the customer's site staff. These should not be overlooked in the preparation of operating cost estimates as they can be significant at sites which do not already have on-site staff available to undertake the necessary duties. Equipment suppliers will provide details of the attendance required on each CHP system component.

A final point to remember where IC engines are concerned is lubricating oil. A 1 MW IC engine will consume upwards of 4,000 litres of lubricating oil per annum, when run continuously. The supply of this oil may or may not be included within the maintenance contract for the engine, so the relevant cost must be allowed for as a separate item, where necessary.

4.9 Estimation of Capital Costs

Approximate installation costs for complete CHP installations abound. Indeed, figures for IC engine and gas turbine based plants are given at the end of chapter 2 of his book. Such figures are, however, far too generalised to be of any practical value in anything but the preliminary evaluation of CHP scheme potential.

To generate a capital cost estimate that is of sufficient reliability and accuracy to be used for the financial appraisal of a scheme, there is no alternative to getting budget quotations for the major system components and compiling an estimate of installation costs based on the work that will be needed at the actual site in question.

If this approach is not taken there is a serious risk that either:

a) A promising scheme may fail to secure finance due to a significant over-estimate of capital costs.

or:

b) A poor scheme may be developed to detailed design stage and even possibly to the point of implementation and then have to be scrapped due to a serious initial under-estimate of capital costs.

The sections which follow provide advice on where and how to get prices and give guidance on how to avoid the potential pitfalls.

4.9.1 Major System Components

Budget quotations for the supply, delivery to site and commissioning of the major system components should be obtained from the manufacturers approached for the performance data required for section 4.7. Even at this budget pricing stage, great care should be taken to ensure that the manufacturer has quoted for all the items you believe he has quoted for. Potential areas of confusion include:

• Does the price for a packaged CHP plant include heat dumping equipment?
• Exactly what automatic controls, electrical protection and switchgear is included in the price for the supply of a generating set?
• Has the waste heat boiler manufacturer included in his price for the supply of the required boost burner and associated controls?
• Exactly what is included with the supply of an acoustic enclosure in the way of attenuators, ventilation fans etc.?
• Does the packaged gas compression plant price include receiver vessels, safety valves and other safety devices and all necessary automatic controls?
• Has delivery, commissioning and testing been included in the price of all components?
• What warranties are provided within the price quoted for each component?

4.9.2 Mechanical Engineering Works

The mechanical engineering works associated with a cogeneration installation are no more demanding than those associated with the installation of any major item of plant. The particular issues that need to be addressed for a CHP scheme include:

Materials   The use of the correct materials for exhaust gas ductwork in particular.

Flexible Connections   Where IC engines are used, due allowance for flexible arrangements to connect all services to the engine must be made to minimise the transmission of vibration.

Sound Attenuation   Adequate allowance must be made for the sound attenuators needed to control the noise emissions from engines.

Air Filtration   In the case of gas turbines the necessary air filtration equipment, which may be automated, can be expensive.

Fuel Preparation   Any necessary fuel filtering, centrifuging and drying equipment must be allowed for.

4.9.3 Electrical Engineering Works

As has been discussed in section 3.8.1, the introduction of generating capacity to a site increases the fault levels that may occur at the site and also on the external supply network. The cost of any necessary reinforcement work can be substantial, particularly where HV systems are concerned. For this reason, it is vital that assistance with the estimation of the cost of the likely electrical engineering work is sought from an expert in the field at this early stage in scheme development. The requirement to spend several hundred thousand pounds on external supply network and site system reinforcement for a scheme, might well swing the economic argument against cogeneration.

4.9.4 Civil Engineering Works

In common with the mechanical works, the civil engineering works required for a cogeneration installation are no more complex than those required for other major items of plant. The particular issues that have to be addressed for a CHP scheme include:

Foundations   Advice should be sought from equipment manufacturers. For generating sets, torsional forces arising as a result of fault conditions will need to be taken into consideration. Finally, for IC engines, vibration isolation arrangements will need to be incorporated into the installation.

Flues   Where possible, separate flues should be provided for each engine of a CHP plant (in some countries, regulations require separate flues). The flues have to be sized to satisfy the back pressure criteria of the engines concerned. Advice should be sought from engine manufacturers on the flue arrangements that will be required.

Price estimates for the necessary works should be compiled in the normal fashion, remembering that the extent of the civil engineering works for a CHP retro-fit project are limited and so unit prices may be higher than would normally be expected. In the pricing of the civil engineering works, due allowance should be made for the removal and reinstatement of existing walls or floors which may be needed to enable the items of plant, in particular the generating set and the waste heat boiler, to be installed.

4.10 The Iterative Process of System Selection

4.10.1 Scheme Optimisation

On a number of occasions in this chapter reference has been made to the iterative nature of scheme development where CHP projects are concerned. To optimise the financial performance of a scheme, under most circumstances, there is no alternative to the sizing, selection, re-sizing and re-selection of system components – engine size being the primary consideration.

As will have been noted from section 4.7, obtaining data and executing calculations for a single set of system components is a time consuming task. Re-executing calculations for a number of different plant sizes does, therefore, demand a significant amount of time and hence cost. For this reason it has become common practice to undersize CHP plants deliberately to make sure that an adequate return on capital is achieved, thereby avoiding the need to carry out optimisation calculations. Indeed the advice given in some quarters is that "It is always better to undersize the engine of a CHP plant, as a second engine can always be installed at a later date if savings permit".

The result of this conservative approach to plant sizing is that many CHP schemes achieve only a proportion of the savings available at the sites concerned. As to the option of adding a second engine at a later date to increase plant size, this will be a non-starter in many instances for the following reasons:

a) For small plants in particular, the capital and maintenance costs associated with two generating sets will be considerably more than the costs associated with one set of the same output. Hence, the addition of a second set to a CHP installation may not be cost effective, whereas the installation of a larger plant in the first instance might have been.

b) Physical space constraints in existing plant rooms may preclude the installation of a second generating set, whereas a single marginally bigger set of twice the output could have been accommodated initially.

c) The mechanical, electrical and civil engineering modifications required to retrofit a second plant may make the proposition uneconomic, whereas the greater works needed for a larger plant in the first place would have added only a marginal sum to the overall installation costs.

There is, therefore, *no alternative to getting the sizing of CHP plants right, from the outset*. With the level of potential savings that can be missed, even where small cogeneration schemes are concerned, there can never be any justification for not making the time and resources available to ensure that plant size is optimised for the site in question.

4.10.2 Optimisation Techniques

For all but the most straightforward of circumstances, the sheer number of calculations that need to be iterated in the process of optimisation will demand that some computer assistance is utilised.

Spreadsheets can be developed to speed up the execution of bin calculations significantly, with the variables that are changed being entered at a single location at the top of the spreadsheet. It will almost certainly be more cost effective, however, to make use of a computer software package designed especially for the purpose. Armed with such a package, the process of iteration can be undertaken quickly, enabling full system optimisation to be achieved. The associated improvement in the financial performance of the project will pay for the cost and use of such software many times over.

Insert panel 3 of this chapter discusses the computer modelling of CHP plant in more detail.

3

THE COMPUTER MODELLING OF CHP PLANT

In the execution of all but the simplest of calculations the use of computers can provide significant time and hence cost advantages. Where a large series of calculations have to be undertaken a number of times, some form of computer assistance may indeed be a practical necessity.

In the design of CHP projects, full scheme optimisation under all but the most straightforward of circumstances can only be undertaken by iteration. Due to the large number of calculations required to predict cost savings/revenue for a single plant configuration, the requirement for iteration means that, in practice, full optimisation of a CHP scheme can only be undertaken with the assistance of a suitable computer programme.

A number of programmes developed specifically to assist in the evaluation and optimisation of CHP projects are now available. In selecting a programme to utilise, consideration should be given to three key elements in the design of the programme:

a) Does the modelling of plant fully and accurately account for changes in equipment performance due to variations both in loading and in outside air conditions?

b) For how many separate time periods are plant performance calculations undertaken by the programme to enable annual savings/revenue predictions to be compiled?

c) What flexibility is provided for the input of complex electricity purchase tariffs?

Item a) above is important in a situation where plant is to be part loaded for a significant number of hours in the year. It is also extremely important where a gas turbine is to be used as the engine of the CHP plant. Item b) will determine the level of accuracy that can be achieved by the programme where unusual and unrepetitive demand patterns are concerned. Finally, item c) will be relevant for sites where complex seasonal time of day tariffs apply to the supply of electricity.

Whilst not necessary for the prediction of CHP scheme performance in many situations, an hour-by-hour approach to modelling provides the flexibility to predict savings adequately under any set of circumstances. For this reason, the hour-by-hour modelling approach is recommended.

5

PROJECT DEVELOPMENT
and
MAKING THE CASE

Perhaps more than any other engineering services project, the results achieved by a combined heat and power scheme are dependent upon the effort and skill that is applied to the design of the installation. In common with all projects, however, a CHP scheme will never leave the drawing board if senior management can't be persuaded of the merits of the proposal.

PROJECT DEVELOPMENT AND MAKING THE CASE

5.1 Generating Set Selection

Initial guidance on the choice of technology and the sizing of output has been given in chapter 4. In this section, the question of manufacturer and model selection is examined. The various issues that need to be addressed are discussed in the paragraphs which follow.

5.1.1 Technical Issues

High Temperature IC Engines   In the selection of an IC engine, perhaps the most important technical issue concerns the operating temperature of the engine jacket and hence the temperature at which heat can be recovered from the jacket cooling water. In terms of power generating efficiency and capital costs, temperatures below 100°C are preferable. In some instances, however, the heating application served will require that heat is recovered from the engine jacket at temperatures in excess of 100°C, as the quantities of heat recoverable from exhaust gases alone would be insufficient to meet the demand for low pressure steam or medium temperature hot water. In such cases, a high temperature engine will need to be chosen.

Engines modified for high temperature operation are made by at least one manufacturer especially for the CHP market. Using this type of engine and by passing water sequentially through the engine jacket and exhaust gas heat exchangers, secondary temperatures in excess of 130°C can be produced. It should be remembered that, even for high temperature engines, the lubricating oil has to be maintained below 90°C to ensure correct lubrication and reasonable oil life. The consequence of this is that 10 to 15% of the recoverable heat from a high temperature engine is still only available at around 80°C.

Nevertheless, where medium temperature hot water or low pressure steam is required at a site, the use of a high temperature IC engine can be vital to the economics of the proposed scheme.

Nearly all IC engine designs were originally conceived with engine jacket water temperatures of less than 100°C in mind. The selection of a supposedly high temperature engine should, therefore, be made with particular caution. Only select a make and model, or model derivative, which has a proven track record at high temperature operation.

Turbo-charged IC Engines   The pros and cons of turbo-charged engines in relation to CHP have already been discussed in section 3.1.4. As most IC engines suitable for CHP were originally designed for straightforward power generation, the capital cost benefits of turbo-charging mean that many of the engines used in packaged generating sets are now fitted with turbo-chargers. In practice, therefore, a high proportion of packaged CHP plants are now built around turbo-charged engines.

Decomposition Gases   The use of decomposition gases and the associated problems have been discussed in chapter 3. Where an engine is to be required to burn such fuels, only manufacturers with first hand experience of the practical difficulties involved should be considered. In addition, where possible, an engine model or model derivative with a proven track record operating on the particular fuel concerned should be chosen.

5.1.2 Energy Costs and Savings

Engine Performance   In the selection of one make of engine over another for the purpose of base load electricity generation, power generating efficiency will be the most important performance issue. In the case of combined heat and power, however, recoverable heat must also be considered.

To compare the overall energy cost savings/revenues for two engines the relevant parts of each bin calculation undertaken for the scheme cost benefit analysis will need to be repeated. Where the analysis work is undertaken with the assistance of computer modelling, it is a relatively quick and easy task to input the different performance characteristics of the engines under consideration, re-run the model and then compare the overall energy savings/revenues that would be achieved by each machine. In cases where hand calculations are used, the process will take longer but cannot be avoided if the correct decision on engine selection is to be made.

In the choice of any engine that is to serve a

base load generating function it must always be remembered that, over the lifetime of the plant, capital cost will be a mere fraction of operating expenditure. For example, the capital cost of a 1 MWe IC engine based generating set (including auxiliaries but excluding installation) would be in the region of £300,000 whilst operating expenditure over 10 years could be in excess of £3,000,000. In the case of such a plant, a 1% difference in brake thermal efficiency between two models of engine might give a £70,000 difference in fuel costs over 10 years. The dominance of operating expenditure over initial capital cost for all sizes and technologies of engine plant is clearly shown by the figures given in tables 3.1 (b), 3.2(b) and 3.3(c) of chapter 3.

Recoverable Heat In comparing the performance of complete CHP packages, it is important to recognise that figures given in manufacturer's catalogues for recoverable heat assume certain heat exchange temperatures and mass flow rates. In comparing the heat recovery performance of CHP packages it is important, therefore, that the figures obtained from manufacturers relate to the actual mass flow rates and temperatures that will prevail at the site in question. Table 4.7(b) of chapter 4 details the information that will need to be supplied to the manufacturers.

Availability The number of hours that a CHP plant is able to operate per annum will obviously have a major impact on the energy cost savings/revenue achieved by the scheme. Engines that are reliable and have been designed to need the minimum of maintenance will obviously achieve a greater number of running hours per annum. Naturally, such machines are usually not the cheapest available. It will, therefore, be necessary to weigh the additional capital costs associated with a high quality engine against the higher level of savings/revenue that is likely to be achieved.

The rationale behind the recommendation that manufacturer's guaranteed availability figures should be used to correct savings predictions for availability has been given in section 4.7.6. Remember that the calculation basis for an availability figure must be known before that figure can be applied to any savings estimate. Manufacturers should be requested to provide figures for availability that include downtime as a result of the execution of both unplanned and planned maintenance work.

5.1.3 Maintenance Costs
The maintenance costs associated with a CHP installation can be equivalent to more than 30% of the value of the energy savings achieved by the scheme. More than 95% of that annual expenditure will be for the maintenance of the engine of the plant. It will be realised, therefore, that a comparison of maintenance costs between engines will form an important part of the selection process.

As is emphasised in chapters 2 and 8, the maintenance of engines is a specialist activity, which is unlikely to be undertaken cost effectively using in-house staff. Quotations for all-inclusive parts and labour maintenance packages will, therefore, be required for the engines under consideration. Section 4.8.2 of chapter 4 gives guidance on how to make sure that 'like for like' contracts are compared. The quotations should be obtained directly from the engine manufacturers or main distributors in the country concerned, as they will be in the best position to provide the high quality maintenance service that will be required. Finally, all agreements offered should be for a 10 year period to ensure that the charges quoted are representative of maintenance costs in the medium to long term.

5.1.4 Service
The importance of the standard of maintenance received by an engine to the overall performance of a CHP plant is discussed in chapter 8. It is, therefore, necessary to get an independent view of the actual quality of the maintenance provided by the engine suppliers under consideration. To achieve this, contact will need to be made with a minimum of 3 users of the maintenance service, preferably in the same geographical area as the site where the cogeneration system is to be installed.

Discussions over the telephone with the users may be adequate but a great deal more information can be gained from a site visit, without the manufacturer being present of course. Questions should be asked regarding the operation of the engine, its reliability, any problems encountered, speed of response from the supplier, the technical expertise and experience of personnel sent to site and the general quality of the service provided.

5.1.5 Financial Security of the Supplier

A 10 year all-inclusive service agreement and warranty for an engine is of no value if the manufacturer ceases to trade. Investigations will, therefore, be required to ascertain the financial security of the companies being considered and their commitment to this particular market. It should be remembered that being part of a large group of companies is no guarantee that financial support for the company will be forthcoming at a time of need. For large projects it may, therefore, be necessary to obtain parent company guarantees in relation to long term maintenance and warranty contracts.

5.2 Ancillary Plant Selection

Preliminary guidance on the items required and the technologies to be used has already been given in chapter 4. It is, nevertheless, worth repeating the advice given that where a suitable CHP package can be sourced from a single supplier this is likely to prove the best procurement route. For larger plants, where the individual components of the system have to be selected, consideration should be given to the issues discussed in the paragraphs which follow.

5.2.1 Boilers and Heat Recovery Equipment

Liquid to Liquid Heat Exchangers Where a close approach temperature between the two fluid streams is not necessary, shell and tube heat exchangers should be utilised. A shell and tube heat exchanger will generally have a lower pressure drop than a plate heat exchanger designed to transfer the same heat rate, as shell and tube exchangers are designed with greater temperature differentials in mind. In addition, shell and tube units are less susceptible than plate heat exchangers to fouling, as the cross sectional area of the waterways is so much greater.

In cases where an approach temperature of a few degrees Celsius is required a plate heat exchanger will be needed. The additional cost of a stainless steel plate heat exchanger will usu-

ally be justified by the greater service life achieved.

Whether shell and tube or plate heat exchangers are selected, the design chosen should facilitate easy cleaning as fouling and the subsequent increase in temperature drop across the heat exchanger will result in a fall off of CHP plant performance. Indeed, in severe cases this may also lead to unplanned IC engine shut downs as a result of high engine jacket water temperature trips. In the selection and sizing of all heat exchangers, realistic 'fouling factors' should be allowed for, based on the water circuit conditions likely to prevail. Advice should be sought from heat exchanger manufacturers on the issue of fouling factors.

Gas to Liquid Heat Exchangers For all but the very largest of IC engines, vertical, in-line exhaust heat recovery units probably provide the best solution in terms of cost and space. Where steam is to be generated, the in-line unit can be manufactured to include a steam space. Alternatively, a remote steam separation vessel can be provided, as illustrated in figure 5.3(b). Finally, for retrofit CHP schemes it is not unusual for an existing boiler to be modified for use as the steam separator for an exhaust gas heat recovery unit.

In the case of gas turbines, a substantial waste heat boiler is required, usually of the water tube type. Due to the relatively low gas temperatures and high volume flow rates in comparison to a normal, fired boiler, waste heat boilers tend to be far larger than conventional boilers. In retrofit projects choice of design may, therefore, be dictated to a certain extent by physical space constraints. In such cases, a vertical design with the feedwater economiser mounted above the main boiler may be the optimum choice where steam is to be generated.

Whether an in-line heat recovery unit or a waste heat boiler is required, the optimum sizing of the unit to give the best return on capital employed should be explored with at least two manufacturers. The minimum information that will need to be given to each manufacturer is detailed in table 4.7(d) of chapter 4.

In common with liquid to liquid heat exchangers, fouling will lead to a significant degradation of heat recovery performance. The problem is particularly acute when fuel oil is burned in the engine of the CHP plant, due to the deposition of soot on the gas side of the heat exchanger. Hence, the chosen design should facilitate the regular cleaning of both the gas and water side heat exchange surfaces. Realistic fouling factors must, of course, be allowed for in the design of exhaust gas heat exchangers, based on the water circuit cleanliness and engine fuel to be used at the site in question.

Where possible, the company that actually conceived the original design of the heat recovery unit or waste heat boiler should be approached directly. This advice stems from experience in the UK, where there have been cases of units manufactured under licence which have not performed correctly due to poor attention to detail resulting from a lack of detailed knowledge of the design and operation of the product. Case histories and contact names of customers should, of course, be obtained.

Finally, this is an appropriate point to repeat the advice given in chapter 3 regarding the use of decomposition gases. The use of sewage gas, in particular, will result in the creation of unusual and particularly corrosive products in exhaust gases. For this reason, only manufacturers of heat recovery equipment who have experience of designing units for use on decomposition gas fired engines should be considered. Customer references will need to be obtained.

Boost Burners   In the case of gas turbines, the boost firing of exhaust gases does not require the introduction of make up air for combustion. Burners located in the gas stream from a gas turbine are, therefore, continuously exposed to the full engine exit gas temperatures of around 500°C. For this reason, burner designs for this application must incorporate some means of cooling certain components of the burner to prevent deterioration through overheating.

Two systems are generally in use. The first uses water diverted from the waste heat boiler to pass through water tubes attached to the burner. The second makes use of air, taken from the boiler room utilising a forced draft system, which is fed to critical components of the burner assembly such as flame detectors.

Whether the boost burner is to fire the exhaust gas of a gas turbine or an IC engine, it is crucial that the design proposed has a proven track record in this application over a minimum of 5

years, as the environment in which the boost burner has to operate is so hostile. The best approach is to allow the manufacturer of each waste heat boiler under consideration to select and propose a make and model of burner that will suit his particular design of boiler. The minimum information that will need to be given to the manufacturers is listed in table 4.7(e) of chapter 4.

Condensing Exhaust Gas Heat Exchangers   In an effort to improve the heat recovery performance of gas fired cogeneration schemes, it is not uncommon to come across the use of condensing exhaust gas heat exchangers. As has been mentioned in chapter 3, however, combustion in engines takes place at far higher temperatures than combustion in boilers with the result that, even when natural gas is being burned, potentially corrosive exhaust products are created in the exhaust gases. Operating experience with some small scale gas engined CHP schemes in the UK has shown that the use of condensing exhaust gas heat recovery units can lead to rapid failure of the units as a result of corrosion.

For this reason, it is recommended that condensing units are made from special, high corrosion resistance stainless steel or the exposed metal surfaces are coated with a corrosion resistant material. Equally, the downstream exhaust ductwork and flue must be made from stainless steel or similarly protected with corrosion resistant material.

5.2.2 Heat Rejection

As has been discussed in chapter 3, in all but the smallest of cogeneration systems, heat rejection facilities must be provided to allow the engine of a CHP plant to be kept operational when the demand for heat from hot water circuits is less than the rate at which heat needs to be rejected from the engine.

For small IC engines and for the lubricating oil cooling systems of gas and steam turbines, direct air blast cooling towers are utilised. These will usually be sized and supplied by the engine manufacturer. For large IC engines, evaporative cooling towers provide the most cost effective solution in terms of fan energy. The additional operating and maintenance costs associated with the use of wet towers will, however, need to be carefully considered when making the choice between dry and evaporative towers. Typical hydraulic arrangements are illustrated in figures 5.3(a), 5.3(b) and 5.3(c).

In the case of exhaust gas heat recovery, there may also be circumstances where the demand for heat may be less than the rate at which heat is being recovered by the exhaust gas heat exchanger. Where steam is being raised, a short-term fall in heat demand can be accommodated by continuing to recover heat at the same rate and allowing the steam pressure in the separator, the on-line conventional boilers and the distribution pipework to rise within the pressure limitations of the system.

Once the pressure limitations of a steam system are reached or where exhaust gas heat is being recovered to hot water, automatic by-pass arrangements will be required to duct the engine exhaust gases around the heat recovery unit. The necessary arrangements are also illustrated in figures 5.3(a), 5.3(b) and 5.3(c). As an alternative to exhaust gas by-pass, cooling towers can be installed for the rejection of excess recovered heat. Such arrangements are, however, usually more expensive in terms of both capital and operating costs and are, therefore, not recommended.

5.2.3 Intakes and Exhausts

Pressure Drop Maximum pressure drop criteria for engines must, of course, be met by the intake and exhaust systems. For IC engines, in a limited number of instances, this may necessitate the use of an induced draft fan system.

For gas turbines, the decision on design pressure drops will be made on the basis of the performance benefits gained versus increased capital costs for larger filters, exhaust gas heat exchangers and silencers. The manufacturer of the selected gas turbine will be able to provide performance figures for a range of intake and exhaust pressure drops to enable an evaluation of the operating cost benefits to be undertaken. Advice can also be sought from engine manufacturers on the designs that have proved cost effective in other, similar, projects. Remember, however, that the use of a certain design and size of component in another project is no guarantee that optimisation of selection has been undertaken.

Table 5.2(a) *Guidance on Optimum Intake and Exhaust Pressure Drops for Gas Turbines*

System	Suggested Range for Total Pressure Drop – kPa
Intake	0.5 to 1.0
Exhaust	1.0 to 2.0

Where a computer programme has been used to assist with scheme evaluation, the consideration of a number of possible intake and exhaust pressure drops is a relatively straightforward task. In instances where hand calculations have been used, it will be necessary to repeat a limited number of the bin calculations to evaluate the cost/benefit of lower pressure drops. It must be emphasised that while this may take additional man time at this stage in project development, the advantages in terms of gas turbine power output can be significant. From section 3.5.2, we will be reminded that each 1 kPa of pressure drop in the intake to a gas turbine reduces maximum power output by about 2%. Guidance on the figures to aim for is given in table 5.2(a).

Engine Intake Air Filtration As has been discussed in chapter 3, the cleanliness of the air supply to an engine will have a significant impact on engine performance. The major benefits of good filtration for IC engines are reduced engine wear and associated reductions in maintenance costs. In the case of gas turbines, poor intake air filtration will, in particular, lead to rapid compressor fouling. As a consequence of this, either down time and maintenance expenditure will have to be increased to keep the compressor blades clean or maximum engine output power and generating efficiency will decline. Where corrective measures are not taken, it would not be unusual to see a reduction in gas turbine maximum power output of 10% after 10,000 or so hours of operation. The issue of gas turbine compressor fouling is discussed in detail in insert panel 1 of chapter 8.

The final design of the intake air filtration arrangements for a gas turbine will, naturally, be dependent upon the particular requirements of the machine concerned but will also need to take into consideration the prevalent air quality at the location of the installation. An offshore scheme, for example, will need to incorporate adequate spray and mist eliminators to prevent the carryover of salt water droplets, as deposits of sodium chloride on gas turbine compressor blades can lead to serious operating difficulties. On land, it is important to locate air intakes as far as possible above ground level to reduce the dust load on filtration systems.

As air filtration is so crucial to the maintenance of engine performance for gas turbines, particular care needs to be taken in balancing capital costs against future operating expenditure. At locations where air quality is poor, additional investment in automatic self-cleaning filters may well prove cost effective.

Like most decisions to be taken in the design of a CHP scheme a balance must be struck between filtration efficacy and intake air pressure drop. To meet filtration requirements for gas turbines, whilst maintaining intake air pressure drop within the guidelines given in table 5.2(a), it will be necessary to specify a filtration area greater than the intake air duct cross sectional area required.

Noise Control   Many acoustic enclosures come fitted with ventilation intake and exhaust sound attenuators. Where this is not the case, calculations will need to be undertaken to determine whether or not there is a requirement to fit attenuators in the enclosure ventilation ducts.

In the case of gas turbines the greatest emissions of noise do not emanate from the engine casing but from the engine air inlet and the exhaust. For a 5 MW gas turbine, for example, typical sound levels from the engine casing (no acoustic enclosure), air inlet and exhaust would be 90, 120 and 105 dB(A) respectively, at 15m. The most effective method of reducing noise disturbance from these sources is to turn the intake and exhaust openings upwards and hence away from the potential listener. Where there are non-industrial buildings close by, however, the use of in-duct sound attenuators will almost certainly be required.

It should be remembered that due to the high rotational speeds employed in gas turbines, the noise produced is greatest at higher frequencies. For example, the sound level at 15m from the air inlet of a typical gas turbine might be just

84 dB for the octave band centred on 250 Hz, whilst at 8,000 Hz it might be 112 dB. In the selection of sound attenuators close attention must, therefore, be given to the individual octave band noise emissions from the selected engine.

Under-provision of sound attenuation can lead to the need for expensive modifications to a project at a later date, along with loss of service or the creation of disturbance to surrounding people. Expert help from a suitably qualified consultant or the manufacturer of sound attenuating equipment should, therefore, always be sought. Particular attention will need to be given to the issue of noise control where engine rooms are located near to residential accommodation.

Finally, the need to fit sound attenuators in exhaust heat recovery by-pass ducts should not be overlooked.

Materials   Calculations may show that the insulation of a flue, both inside the engine room and outside, will be necessary if minimum exhaust gas temperatures are to be maintained up to the point of exit from the flue. All potential cold spots at the location of fittings, supports etc. along the length of the flue must also be protected against corrosion through the judicious selection of materials or the application of corrosion resistant coatings.

It might be thought that when the necessary precautions have been taken to ensure that bulk condensation is avoided, ordinary steel ductwork could be used for engine exhaust installations. Operating experience in the UK and elsewhere has, however, shown that when ordinary mild steel is used, sections of the exhaust system, particularly bends, have to be replaced every two or three years as a result of corrosion and erosion. It is, therefore, strongly recommended that twin wall stainless steel flues (and flue liners where brick or concrete chimneys are to be used) are specified for engine exhaust installations.

Where fuel oil is to be burned by an engine on a regular basis, even standard stainless steel will not be adequate. In this case, due to the sulphur content of the fuel and the resulting production of sulphurous acid in the exhaust gases, special austenitic type stainless steel will need to be used.

Where condensation is actually encouraged, through the use of a condensing exhaust gas heat recovery unit, the precautions mentioned in section 5.2.1 need to be taken.

As regards engine room ventilation and gas turbine intake ductwork, there are no special requirements for the materials to be used. To control noise levels in the engine room, however, gas turbine intake ductwork will probably need to be lined with acoustically absorbent material.

5.2.4 Gas Compressors, Metering and Fuel Oil Supply Equipment

Gas Compressors In the selection of gas compressors, the fact that a manufacturer has supplied equipment for successful CHP installations, whilst important, is not the crucial issue. Experience in the UK has shown that where problems with gas compressors do arise, they usually concern the overall gas supply system or the automatic controls utilised rather than the compressors themselves. What is important, therefore, is to find compressor suppliers who have a proven track record in the design and assembly of complete gas supply packages.

Metering There is some discussion in the industry as to the suitability of turbine type gas meters when used in conjunction with IC engines, due to reading inaccuracies caused by the pulsating flow associated with reciprocating engines. This potential difficulty can be overcome by locating meters at a distance from engines or alternatively by installing a small snubber vessel between the meter and the engine. Advice can usually be obtained from the local gas supply utility.

Fuel Oil Supply Equipment The treatment required for fuel oils has been mentioned briefly in section 4.9.2 in relation to costs. In cases where heavy fuel oils are to be used, the conditioning of oil prior to supply to the engine will include heating, centrifuging and filtering. Fuel oil specification at the point of supply to the engine must, of course, be approved by the engine manufacturer as it will have a fundamental impact on the performance of the plant and the associated maintenance costs. In fact, it is usual for the necessary fuel treatment equipment to be purchased as a package from the engine supplier.

When use is to be made of existing supply arrangements, fuel pumping, day service tank capacity, fuel heating (in the case of heavy fuel oil) and fire safety equipment will all need to be reviewed in relation to the planned CHP installation.

5.2.5 Alternators
Guidance on the choice of generating voltage has been given in section 4.6.6 of chapter 4. As alternators will be sourced as part of a packaged generating set, selection of make and model is normally undertaken by the set packager. Selection criteria for alternators are, thus, not given here.

5.2.6 Electrical Switchgear and Protection
As has been emphasised in chapter 4, this is one area in the development of a CHP scheme where specialist advice must be sought, particularly where high voltage installations are concerned. A discussion of the pertinent criteria for the selection of the necessary switchgear and protective devices, therefore, falls outside the scope of this book.

5.2.7 Automatic Control and Monitoring
As has been discussed in section 3.9.1, the consequences of the failure of the automatic controls serving a generator can be serious. Equipment must, therefore, be chosen that has the necessary reliability to achieve the high security of service that is required. In addition, unauthorised changes to operating control strategies could have major implications in terms of security of service and energy cost savings/revenue. For these reasons, the extension of an existing building management system at a site to take control of a CHP plant is not a practical proposition. It is usual practice to select the control system that is offered by the supplier of the chosen generating set. This system can then be designed and configured to meet the requirements of the complete CHP installation.

CHP plant operating strategies are discussed in detail in chapter 8. The following issues should, however, be born in mind at this stage of project development:

• Full co-ordination of control with existing mechanical services, particularly boiler plant, will be required.

- Consideration should be given to interfacing the new CHP plant control and monitoring system with the existing control systems at a site. This is discussed in section 5.3.4.
- The monitoring and data logging facilities provided by the system will have to be adequate for the commissioning and testing of the CHP plant, in addition to meeting the on-going operational and maintenance requirements.
- The system will have to have the monitoring, data input and programming sophistication necessary for it to be able to deliver the operating control strategies specified for the CHP plant.

5.2.8 General

Procurement options for CHP schemes are reviewed in chapter 6. It is important, however, to emphasise here that the adoption of an *intelligent, 'value for money' approach* is of particular importance to the selection of components for a CHP scheme for two reasons, which are:

a) The correct matching of the technical performance of the individual components of a cogeneration scheme is vital to the successful operation of the plant and to the achievement of the optimum return on investment. It is, thus, highly unlikely that the optimum plant for a particular site will be selected if a lowest capital cost procurement route is followed.

b) The fundamental rationale behind a decision to go for combined heat and power is one of investment of additional capital resources to achieve future revenue savings. As the life cycle operating costs of a cogeneration plant dwarf the capital costs, a lowest capital cost selection criterion for the actual components of the CHP scheme simply does not make sense.

5.3 Interfacing and Installation Works–Mechanical

Up to this point in the book, consideration of the existing mechanical services at a site has been limited to discussions on heat demands and heat transfer. The way in which a cogeneration system is integrated into an existing installation will, however, be a major contributing factor to the success or failure of the project. Mechanical interfacing and the associated installation works are discussed in the paragraphs that follow.

5.3.1 Recovery of Heat to Water Circuits

Hydraulic Arrangements The hydraulic arrangements chosen for a cogeneration system have to meet one simple criterion – that they achieve maximum heat recovery from the CHP plant, within the constraints of practical site considerations. In terms of IC engine jacket water and lubricating oil cooling this means:

a) Taking the maximum possible flow rate of heating circuit water through the CHP heat exchanger(s).

b) Maximising the heat exchange temperature differential by routing return water through the CHP heat exchanger(s) before it has been heated by any other means.

c) Achieving maximum running hours for engines which do not have alternative heat rejection arrangements, by ensuring that the CHP plant operates in preference to boiler plant.

d) Ensuring that engine heat rejection equipment only operates when insufficient heat is being removed from the engine through heat recovery.

To satisfy these requirements, jacket water and oil cooling heat exchangers should always be connected in-line in the return pipework of the heating circuit to be served. To get access to the maximum possible volume flow rate and hence realise the maximum potential for heat recovery, the heat exchangers have to be connected in the common return piping for the system which, naturally, will be located close to the boiler plant serving the circuits in question.

CHP units should not be connected in parallel with existing boilers as, in this configuration, reliance has to placed on the correct operation of automatic controls to ensure that boilers only fire when the CHP plant is not able to satisfy the demand for heat. In addition, unless motorised boiler isolation valves are installed, only

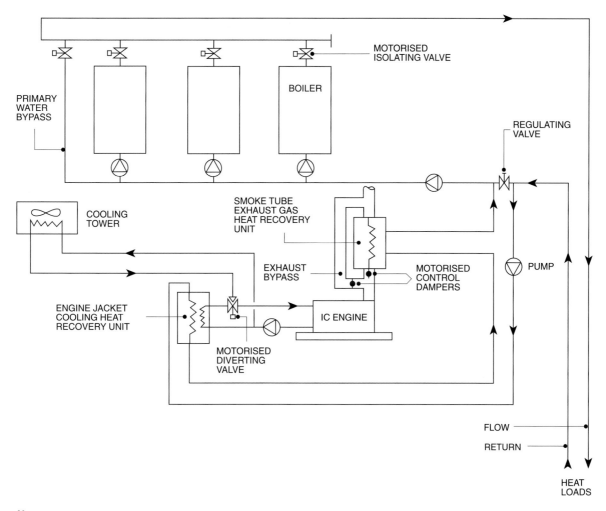

Notes

1. Engine lubricating oil and aftercooler (where fitted) heat rejection arrangements not shown for simplicity.

Figure 5.3(a) *Simplified Diagram of Typical Hydraulic Arrangements for the Recovery of Heat – IC Engine Generating Hot Water*

a proportion of return water will pass through the CHP heat exchangers, thereby reducing heat recovery performance. The in-line connection configuration inherently avoids these potential difficulties.

Figure 5.3(a) illustrates the recommended hydraulic arrangements for the recovery of heat from IC engines to water circuits.

CHP Plant Location As was illustrated in table 2.3(a) earlier in the book, hot water/steam pipework is more expensive to install than electrical cabling in terms of capital costs, though not necessarily in relation to the value of energy

transported. It is, therefore, usual though not vital for the generating set of the cogeneration system to be located adjacent to the other heat producing plant at a site. In fact, for smaller systems, it is not uncommon for generating sets and boilers to be housed in the same room.

Existing Water Circuits A CHP system cannot, of course, compensate for the poor hydraulics of an existing installation. Where retro-fit projects are concerned it is important that the opportunity is taken to review thoroughly the operation of hot water circuits at the site in question, as an integral part of the CHP investigations. Steps should then be taken to eliminate any difficulties identified, as part of the co-generation project installation works.

Re-sizing of Pumps   The installation of a heat exchanger, particularly a plate heat exchanger, may significantly increase the pressure drop in a hydraulic circuit. The checking of pump sizing when a CHP plant is to be retro-fitted into an existing installation must, therefore, not be overlooked.

It will often be necessary to change the pulley sizes on belt driven pumps to accommodate the increased circuit resistance caused by heat exchangers and, in some instances, actual motors and pumps may need to be replaced.

Water Treatment and Cleanliness   The treatment and standards of cleanliness required for engine jacket cooling water are quite different to those generally required for hot water circuits. *Physical separation between the two water streams must, therefore, always be provided through the use of a heat exchanger.*

At sites that have been in operation for a number of years, there may be considerable quantities of suspended particulate matter flowing in the water circuits. Normally this does not cause any particular operating difficulties where standard heating systems are concerned. Heat exchanger fouling is, however, a major cause of loss of performance with CHP systems. Where a CHP plant is to be installed at an existing site, therefore, steps will need to be taken to clean up heating circuits and to install suitable strainers upstream of each heat exchanger. The use of plate heat exchangers demands a particularly high level of water cleanliness, if excessive maintenance expenditure on exchanger cleaning is to be avoided.

5.3.2 Recovery of Heat to Steam Systems

IC Engines   Where steam is to be generated from the exhaust gases of an IC engine, a gas to water heat recovery unit with a remote steam separator is often the most convenient arrangement. Such an arrangement is illustrated in figure 5.3(b). Also shown in the diagram is a 'feedwater economiser'. This is simply a heat exchanger which enables additional heat to be recovered from the engine exhaust gases to the steam separator feedwater. Where a feedwater economiser is used, however, particular care will need to be taken to ensure that minimum exhaust gas temperatures are maintained under all operating conditions. For some installations, this may necessitate the mixing of feedwater with water circulated from the steam separator, using a three port valve arrangement, to maintain a minimum 'on' temperature to the economiser.

Recent experience in the UK has highlighted a problem that can occur with lean burn engines fitted with automatic ignition retardation control. When the control system acts to retard ignition timing to prevent engine pinking, the associated increase in exhaust gas temperature may cause the CHP plant to trip as a result of high pressure in the steam separator of the waste heat recovery system.

Where the use of such an engine is envisaged, care must be taken to ensure that the exhaust gas heat recovery arrangements are designed to accommodate this type of temporary pressure rise without tripping the CHP plant.

It is worth emphasising here, that as only roughly 15% of the recoverable heat from an IC engine is available in the exhaust gases, an IC engined CHP system designed to raise only high pressure steam will almost certainly not be an economic proposition. The use of special high temperature engines, which operate with engine jacket water temperatures adequate to raise low pressure steam, has been discussed in section 5.1.1.

Where the selection of a gas turbine is not possible at sites that utilise high pressure steam, consideration will, therefore, have to be given to using IC engine jacket and lubricating oil heat to serve a hot water circuit directly. At most steam sites it is usual to find a significant proportion of the final distribution of heat being undertaken in the form of hot water, generated in remote steam/hot water calorifiers. The question will be whether the connection of a CHP system heat exchanger directly into one of these remote hot water circuits would prove cost effective, in terms of the additional capital costs involved and the heat recovery potential. The alternative may be actually to locate the generating set adjacent to the remote steam/hot water calorifier and recover all heat from the engine to hot water.

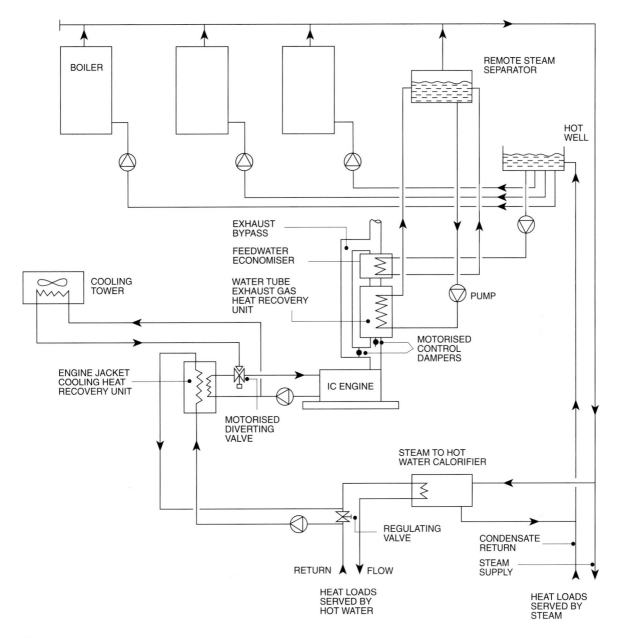

Notes

1. Engine lubricating oil and aftercooler (where fitted) heat rejection arrangements not shown for simplicity.

Figure 5.3(b) *Simplified Diagram of Typical Hydraulic Arrangements for the Recovery of Heat – IC Engine Generating both Hot Water and Steam*

Figure 5.3(b) shows the recovery of heat from an IC engine to both steam and hot water.

Gas Turbines Where a gas turbine is to be used, virtually all the recoverable heat is available at high temperature in the exhaust gases from the engine. The generation of high pressure steam is, thus, not a problem. A suggested hydraulic arrangement is shown in figure 5.3(c). In situations where feedwater temperature is expected to vary, the mixing of feedwater with water circulated from the steam drum may be

141

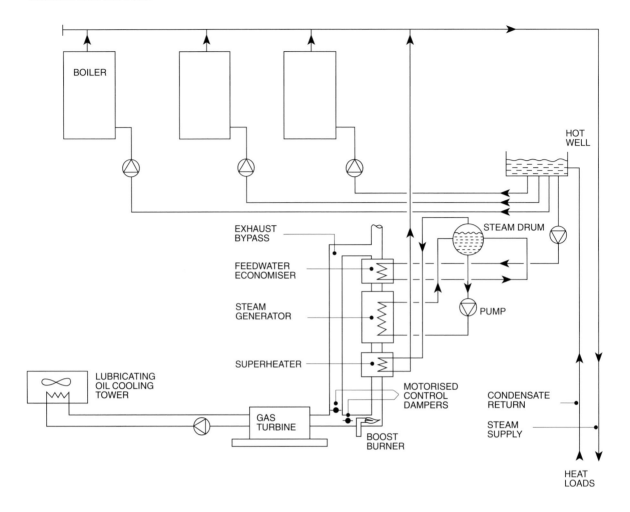

Figure 5.3(c) *Simplified Diagram of Typical Hydraulic Arrangements for the Recovery of Heat – Gas Turbine Generating Steam*

required to ensure that minimum exhaust gas temperatures are maintained. This requirement has been discussed under 'IC engines' above.

For combined cycle installations or where existing site equipment demands it, superheated steam will need to be generated from the gas turbine exhaust gases. This will necessitate the provision of an additional heat exchanger, known as a 'superheater', which is located in the exhaust gas stream before the main steam generator. Dry saturated steam is taken from the steam drum and passed through the superheater before being supplied to the site. The necessary hydraulic arrangements are illustrated diagrammatically in figure 5.3(c).

5.3.3 Gas Supplies
The design of the fuel gas supply installation for a gas turbine requires specialist expertise. Whilst the necessary expertise can be provided by a consultant, it has become common practice to approach compressor suppliers to provide a complete packaged gas supply plant. This approach is recommended here, as companies with a proven track record in the design and supply of complete systems are more likely than consultants to have built up the necessary level of hands-on experience to avoid the pitfalls that are sometimes encountered.

To assist the reader with the selection and subsequent performance monitoring of the compression plant supplier, guidance on the issues which will need to be considered is given here.

Start-up On start up, some gas turbines can be run from rest up to full power in less than 30 seconds. The acceleration and compression of gas to the required delivery velocity and pressure causes significant pressure transients that can affect metering equipment and the supplies to adjacent customers if the necessary features are not designed into the supply system. For the majority of CHP systems, however, there is no requirement to bring the gas turbine rapidly up to full load on start-up. This means that the turbine and gas compressor can be loaded gradually over a period of time, thereby avoiding the problem of pressure transients on start-up.

In those cases where fast start-up and loading is required, it is often preferable to operate a gas turbine on fuel oil initially and then switch to gas after a short delay.

Shut-down Shut-down can be a much more violent occurrence than start-up, as it may have to be executed in under 1 second in the event of a turbine trip. Two problems can occur. If the inertia of the rotary components of the fuel gas compressor is great enough, the compressor will continue to deliver gas into the receiver vessel whilst the valves from the receiver to the gas turbine are shut. Adequate pressure relief arrangements on the vessel are, therefore, required to ensure that design working pressure is never exceeded.

The converse of this eventuality is that high pressure gas from the receiver 'spills' back through the compressor to the low pressure supply. This is of particular concern to gas supply companies, who will insist that a slam shut valve is installed before the compressor to protect metering equipment and to prevent the occurrence of dangerously high pressures in the low pressure distribution network.

Reciprocating Compressors and Resonance Where reciprocating compressors are used, during normal running the reciprocating action of the pistons generates regular pressure pulses which, under certain circumstances, can resonate within the supply system to generate quite violent vibrations.

This problem can be overcome by careful design and sizing of the supply pipework along with the inclusion of a small 'snubber' vessel at the inlet to the compressor.

Compressor Lubricating Oil Carry-over The oil separation arrangements provided for the compression plant must control lubricating oil carry-over to the level that is required by the chosen gas turbine manufacturer.

Automatic Controls Different sets of automatic controls may be used to control the gas compression and the gas turbine plants. Problems of control instability can arise when the two independent sets of controls interact, leading to hunting of the compressor and gas turbine. Such problems can be overcome through the installation of an adequately sized receiver vessel between the compressor and the turbine and the use of a single set of integrated controls to serve both plants.

Compliance with the Requirements of the Gas Supply Utility The packaged supply system will need to incorporate any operational or safety features required by the local gas supply utility. These will vary from site to site and from utility to utility.

General Supply Arrangements Notwithstanding the need for expert assistance, the general arrangement and components of a gas turbine fuel gas supply system are illustrated in figure 5.3(d).

5.3.4 Automatic Control and Monitoring
As has been discussed in section 5.2.7, the control and monitoring system for the CHP installation is unlikely to be of the same make and model as the other control systems in use at a site. For small CHP installations, particularly where no technical staff are based on site, this will not be a problem. For large sites, however, some form of integration between the various systems will be advantageous. Software packages are now available that can be configured to supervise the operation of a wide variety of different proprietary control systems, provided the necessary communication protocols are written to allow the systems to talk to one another.

5.3.5 General

Flexible Connectors Though not required for the purpose of vibration isolation on gas and steam turbines, it is recommended that use is

Figure 5.3(d) *Simplified Diagram of Typical Gas Turbine Fuel Gas Supply Arrangements*

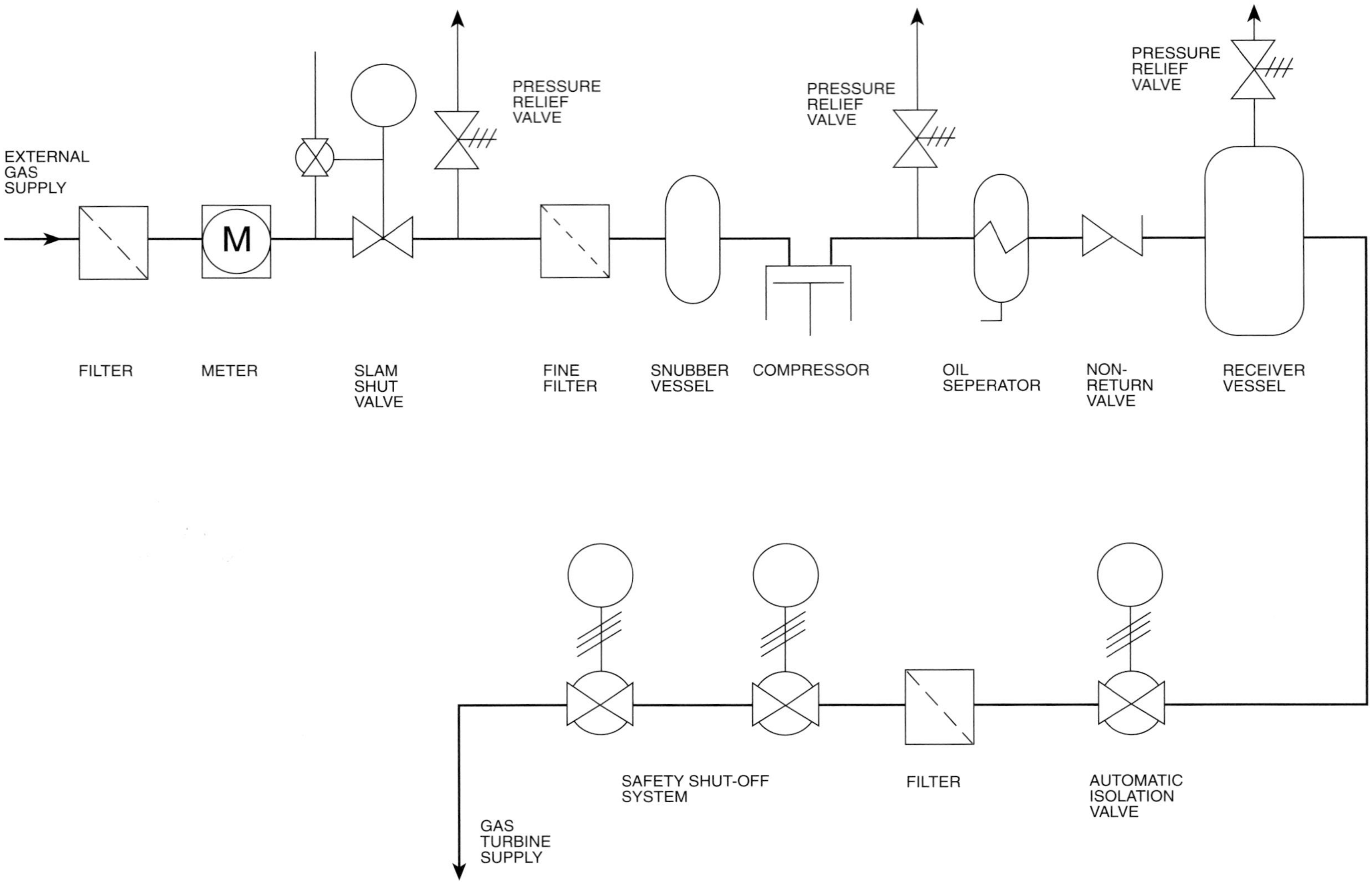

EXTERNAL
GAS
SUPPLY

FILTER

METER

SLAM
SHUT
VALVE

PRESSURE
RELIEF
VALVE

FINE
FILTER

SNUBBER
VESSEL

COMPRESSOR

PRESSURE
RELIEF
VALVE

OIL
SEPERATOR

NON-
RETURN
VALVE

PRESSURE
RELIEF
VALVE

RECEIVER
VESSEL

GAS
TURBINE
SUPPLY

SAFETY SHUT-OFF
SYSTEM

FILTER

AUTOMATIC
ISOLATION
VALVE

144

made of flexible connectors for all mechanical connections to engines, to ensure that excessive force is not applied to any connection flanges. Points to remember when using flexible connectors are:

- Exhaust connectors must be able to withstand the transient high exhaust temperatures that, on occasion, occur with IC engines. Stainless steel connectors are recommended.
- Connectors affixed to IC engines should be mounted such that the principle motion is at right angles to the axis of the connector, i.e. engine vibration does not cause stretching and compression of the connector.
- Connectors should be installed without offsets and with no twists in the connector.

Engine Exhaust Systems The general requirements for exhaust systems, components and materials have been given in sections 3.5 and 5.2.3 of the book. The following additional items should be considered in relation to the detailed design of exhaust installations:

- To minimise noise in the engine room, the distance between engines and exhaust silencers should be kept as short as possible.
- For IC engines, where it is not possible to provide separate flues for each engine (and where applicable legislation or codes of practice do not require separate flues), a 'Y' connection rather than a 'T' connection must be used to join the two exhausts. This will prevent the exhaust pulses from one engine interfering with another.
- For IC engines, exhaust length should be chosen to give an odd multiple of the critical length given by the engine manufacturer.
- Flexible connectors or other arrangements will need to be provided to allow for approximately 10 mm per 1m of exhaust duct length for expansion.
- Horizontal runs must be installed with a fall away from the engine to ensure that, on start-up, condensate does not run back into the engine.
- Drain points connected to pipework running to waste are required at all low points in the exhaust system.
- The pipework used for condensate drains should have an internal diameter of at least

10 mm to prevent clogging and to facilitate regular cleaning. Copper should not be used as it corrodes relatively quickly in the presence of engine exhaust gases. Plastic or galvanised steel pipework is recommended.
- Condensate drains must incorporate water traps with a sufficient head of water to prevent the release of exhaust fumes into engine rooms.
- Stack height must be chosen to satisfy local legislation for the fuel to be burned.
- For IC engines, exhaust ductwork, flexible connectors, silencers and heat recovery units must be designed to withstand the sudden rises of pressure associated with engine backfire. Advice should be sought from the manufacturer of the chosen engine.
- Exhaust ductwork joints should preferably be welded or flanged.

Physical Space for the Maintenance of Plant and Equipment As with all mechanical plant, adequate space must be left around the various components of a cogeneration system to facilitate the maintenance of each item of equipment. Advice from the manufacturer of each component of the plant should be strictly adhered to, as the regular and involved nature of CHP plant maintenance would make inadequate space intolerable. Special attention should be paid to the need to dismantle engines and remove components, to withdraw tube bundles from shell and tube heat exchangers and the need to disassemble plate heat exchangers.

The Use of a Steam Accumulator At sites where there are sharp peaks of demand for steam for relatively short periods of time, the installation of a large steam vessel to accumulate steam can provide significant operating benefits. The idea is that by generating steam in excess of demand outside peak periods using the CHP plant and by storing that heat energy in the accumulator, peak loads can be met without the need to bring alternative heat producing plant on line.

5.4 Interfacing and Installation Works – Electrical

Notwithstanding the previous comments made

in the book regarding the requirement to get expert assistance with the design of the electrical interface for a CHP system, outline guidance on the issues that will need to be considered is given here. This guidance is provided only to assist the reader in the selection and subsequent performance monitoring of the required specialists.

5.4.1 Connection Arrangements

Switching and Protection   The electrical switching and protection arrangements that may be required for a CHP installation have been discussed in some detail in section 3.8. From this discussion, it will be realised that prospective fault levels, generating plant, system voltage and the characteristics of the external supply network all have an impact on the exact protection arrangements required at a particular site.

For small CHP systems of a few hundred kWe output, where asynchronous alternators are to be used, a qualified electrical engineer should be able to devise the necessary switching and protection with the help of this book and the chosen generator supplier. For larger plants and when generation is to be at high voltage, assistance from a specialist in embedded generation will be required.

Standby Operation   At sites where a CHP plant is to provide standby power in the event of external supply failure but where site electrical demand exceeds the capacity of the generating plant, automatic load shedding arrangements will have to be put in place. Load shedding has been discussed in section 3.8.8 of chapter 3.

It should be recognised that at sites where electrical services are not already separated into essential and non-essential categories and wired accordingly, the retro-fit of load shedding is likely to be costly.

5.4.2 Supply Earthing

Neutral earthing of electrical supply systems is undertaken to cause sufficient current to flow to ensure that protection equipment operates effectively in the event of earth faults and to limit the voltages appearing on healthy phases during a fault.

Where a site is supplied at low voltage, the neutral of the supply network will be earthed at the local utility company transformer. It is quite common, therefore, for an earthing terminal to be provided by the utility company at the point of supply to a site. Some external networks employ 'protective multiple earthing (PME)', in which case the neutral of the supply is also connected to earth at the point of supply.

The neutral of a high voltage supply, on the other hand, may be earthed at some remote point on the network a considerable distance from the site supplied. Sites supplied at high voltage are, therefore, required to have their own earthing terminal connected to an earth electrode on-site.

The earthing of supply neutrals at multiple points on a network can lead to problems with the effective operation of earth fault protection and the achievement of adequate discrimination between earth fault protection equipment at different points on the supply network. In addition, multiple neutral earthing may cause the flow of unacceptable harmonic currents in the neutral. For this reason, the connection of the neutral of a site distribution system is avoided, unless the external network employs, and therefore accommodates, protective multiple earthing.

A problem thus arises where local generating plant is required to operate both connected to an external network and in isolation. When islanded from the external network, the site supply system must be effectively connected to earth at some point at the site, whereas under parallel operation local neutral earthing is not permissible.

A number of options are available to overcome this problem and these are discussed in the paragraphs which follow.

Parallel Only Operation   Where the generating plant at a site is only ever to be operated in parallel with an external supply, then no provision need be made for local neutral earthing.

Protective Multiple Earth Networks   Where the external supply network is of the protective multiple earth type, at low voltage it may be acceptable to connect permanently the star points of the alternators at the site to a local earthing electrode which is bonded to the external supply earthing terminal. Should a utility company earthing electrode be located at the point of supply to the site, then use may be made of this electrode in lieu of a site earthing electrode.

Standby Neutral Earthing at Generators With this arrangement, in parallel mode the star points of the site alternators are unconnected and the site supply neutral is connected to earth via the external supply. In the event of external supply disconnection, the neutral point of one of the site alternators is automatically connected to a local earthing electrode using a circuit breaker or contactor. On reconnection of the external supply, the alternator star point is disconnected once more by the automatic opening of the neutral earthing breaker/contactor.

Standby Neutral Earthing using an Earthing Transformer In this system, alternator star points are permanently disconnected from earth. Instead, an earthing transformer (or transformers where there is more than one bus-bar section) is connected between the paralleling bus-bars and a local earthing electrode. Under parallel operation the earthing transformer is isolated from the bus-bars using a circuit breaker. On external supply disconnection, the circuit breaker automatically closes to connect the earthing transformer.

5.4.3 Fault Current and Reinforcement

Prospective Fault Level In the event of a short circuit or earth fault on a system, supply impedance is dramatically reduced leading to the temporary generation of high current flows by generating plant, significantly in excess of peak generating capacity. These high current flows are known as 'fault currents'.

As has been discussed in sections 3.7.1 and 3.7.2, it is synchronous alternators that make the major contribution to the generation of fault current, although asynchronous machines will self-excite and continue to generate at a reduced output for a short period of time as the magnetic flux in the rotor decays. Indeed, induction motors will also act as induction generators making a contribution to fault current when a supply is lost.

At a site with its own generating plant, contributions to fault current will, thus, come from local generators, induction motors and, when the system is connected to an external network, the generators of the utility company. The total current expected to flow at any point on a site distribution system or external network is termed the 'Prospective Fault Level'. The prediction of fault levels is discussed in insert panel 1 of this chapter.

Fault Current Limiting Under some circumstances it may be advantageous to limit the maximum level of current flow under fault conditions. When generating plant is to be retrofitted to an existing site, for example, the site distribution system and/or external supply network may not be able to handle the increased prospective fault levels that would arise should current limiting measured not be applied.

There are essentially two main methods used to limit fault currents. These are discussed in the paragraphs which follow. In rare instances a third option may need to be considered. That is selecting or having manufactured specially constructed alternators which have a low fault current characteristic.

Explosive Fuses Where the installation of embedded generating plant would otherwise cause prospective fault level problems, the installation of 'explosive fuses' at the intake to the site may satisfy some utility companies that fault current on their network will be limited to acceptable levels.

The high speed current limiting characteristics of HRC fuses enable them to interrupt fault currents within half a cycle of fault occurrence. This is much faster than circuit breakers operate. By interrupting faults so quickly, HRC fuses significantly reduce the energy that flows as a result of a fault, thereby providing a greater level of protection to an electrical distribution system. With present HRC fuse designs, however, it is not possible to select a fuse that is rated to pass the full output of a particular generator continuously, but which will also provide half cycle disconnection on the level of fault typically produced by that generator.

This problem can be overcome by paralleling a smaller fuse with a bridge which contains an explosive element. Under normal conditions, current is shared by the fuse and bridge and hence is well within the continuous rating of the fuse. In the event of a fault, the explosive element is detonated by a rate of rise of current detector, all the fault current passes through the fuse and the fuse operates within half a cycle.

Just like an HRC fuse, it is not possible to test the operation of an explosive fuse and hence check it is in working order, on a regular basis. The failure of a fuse to operate in the event of a fault on a low voltage system might result in some damage to local cables and bus-bars. In contrast, the failure of an explosive fuse to operate to protect high voltage cables, bus-bars and switchgear could be catastrophic. For this reason, some utility companies do not view explosive fuses as being a reliable means of fault limiting and hence protecting their network from excessive fault levels.

Finally, experience in the UK has revealed that the induced voltages from high frequency transients can cause the operation of the detonator in some designs of explosive fuses. Where high frequency transients are expected at a site (generated by the inverters of uninterruptible power supply systems, for example) and explosive fuses are to be used, care needs to be taken to select a design of device that is not affected by transients.

Reactors   The most widely used device to limit fault currents from generators is the 'reactor'. A reactor comprises a coil with a low resistance but a high inductance. The device therefore consumes little 'real' power but acts to resist changes to the rate at which electrical current flows. In the event of a fault, therefore, the inductive nature of the reactor coil will impede the rate at which fault current builds up, thereby reducing the maximum fault level experienced on the system concerned.

In terms of normal operation, reactors present a negligible 'real' load on a system and hence cause little 'real' power loss. On the other hand, reactors are designed to have a high inductive impedance. They, therefore, place a significant reactive load on a supply and act to worsen lagging power factor. If the total reactive load at a site exceeds the MVAr rating of the local generating plant, then supply voltage will become depressed. The reason for this is discussed in insert panel 8 of chapter 3.

Where reactors are installed at the intake to a site to protect an external network from excessive fault current, supply voltage depression can be a problem. When the site is largely being supplied by its own generating plant, there will be little import of power via the reactors and hence little reactive load presented by the reactors. Should a generator be taken out of service or the output from individual generators be reduced, under a cogeneration control strategy for example, a substantial level of power will then be imported through the reactors. This may take reactive loading beyond the MVAr capacity for the generating plant, with the result that supply voltage can no longer be maintained above the minimum standards set for the supply.

In the case of the loss of a generator, this potential problem can be eliminated through the use of a number of reactor sets in series, one set for each generator. By-pass breakers installed across each reactor set and interlocked with the relevant generator breakers, then ensure that reactors are only switched into the circuit as and when required.

Where individual generator power output control leads to increased import and hence excessive reactive power loading, auxiliary power factor correction equipment may be required.

Reinforcement   Where a CHP plant is to be installed at an existing site and operated in parallel with a utility company supply, the upgrading of cables, bus-bars and switchgear may become necessary on the site distribution system and/or the external supply network. These remedial works are known as 'reinforcement'. The requirement for reinforcement may arise for a number of reasons, which include:

- Increased full load current flows on certain parts of the system.
- Higher prospective fault levels which cannot satisfactorily be limited using reactors or explosive fuses.
- The need to maintain discrimination with an increased number of protective devices installed in series.

From the above and from previous discussions in the book on fault levels, protection arrangements and discrimination, it will be realised that the investigations needed to determine whether and what reinforcement work is

1

FAULT LEVEL AND STABILITY PREDICTIONS

Prediction of Fault Levels The items of plant that contribute to fault current at a site have been discussed in section 5.4.3. Where site generating plant is only to operate in isolation from a utility company supply, the prediction of fault levels at various points on the distribution system can be undertaken using hand calculations. Computer programmes are, however, available for this work making hand calculations unnecessary, except for the rough checking of the answers provided by the computer.

In the case of parallel operation, both site plant and the plant of the utility company will contribute to fault current. This makes the use of a suitable computer programme a practical necessity when predicting fault levels for systems with embedded generation.

Whether or not a computer package is used, the prediction of fault levels is a job for a specialist.

Generator Stability 'Generator Stability' concerns the maintenance of synchronism between synchronous generators and an external supply. Stability can be considered under two categories:

• Steady State Stability
• Transient Stability

'Steady State Stability' is concerned with the maintenance of synchronism under normal operating conditions. In the case of modern generators with automatic excitation control and fast acting governors, steady state stability is unlikely to present any problems.

'Transient Stability', on the other hand, relates to the regaining of synchronism by a generator after the occurrence of a fault or other unusually large change of load on the system. When a fault occurs on a supply, the output voltage from a local generator will collapse. The power delivered by the alternator of the set will, therefore, fall suddenly, whilst the power delivered by the engine remains constant, initially. The result is that the engine and alternator start to accelerate.

The degree of acceleration that takes place will be dependent upon the magnitude and duration of the fault, the inertia of the engine and alternator (mass of rotating parts), the response time of the engine governor control system and the response time of the alternator excitation control system. With the alternator rotor running at greater than synchronous speed, pole slipping will occur with the associated risk of rotor overheat.

On clearance of the fault, the alternator rotor may almost immediately pull back into synchronism with the external supply, may pull back into synchronism after a few seconds or may continue to pole slip indefinitely.

In terms of transient stability, the most onerous fault on a generator is a 3-phase short circuit. Such a fault is, thus, used as the basis for calculations to determine the maximum fault duration time following which a particular generating set will immediately pull back into synchronism. This time is known as the 'Critical Fault Clearance Time' for the installation and generator concerned. Computer simulation programmes are available to assist with the prediction of critical fault clearance times.

Where possible, generator protection arrangements should be designed and set to achieve the disconnection of short circuit faults in less than the critical fault clearance time for the plant and system under consideration. If this proves impractical, then additional Generator Protection, in the form of phase displacement relays, may need to be installed to detect pole slipping and trip the generator breaker after a set time delay.

required are extensive where a major CHP installation is concerned. It should also be recognised that in situations where reinforcement work is unavoidable, the capital costs involved are likely to be substantial. This is particularly true when the upgrading of utility company networks is required. As this work is undertaken at the expense of the customer, the value of discussing a potential CHP project with the local utility company at an early stage in scheme development, can not be over emphasised. The advantage of a good working relationship with the utility company will soon become apparent when discussions of the protective and reinforcement requirements start. The option of plant procurement through a joint venture with the utility company is discussed in chapter 6. Such an approach can produce significant benefits when it comes to interfacing with the external supply network.

5.4.4 General

Power Factor Correction Equipment Automatic power factor correction equipment makes use of reactive current transformers to generate a control signal to switch in the appropriate number of capacitors. Where a CHP system is both to import and export power, relays will need to be fitted to reverse the contacts on the current transformers when export takes place, to allow the power factor correction equipment to continue to function.

In addition, generator protection arrangements must be designed to ensure that an alternator cannot be isolated from the main electrical distribution system at a site, whilst remaining connected to power factor correction capacitors. This is to prevent the generation of large transient overvoltages that would otherwise occur in the isolated circuit.

Fire Protection A number of the components of a CHP plant present a significant fire risk. Where supplied inside an acoustic enclosure, a generating set will usually be provided with an automatic fire detection and extinguishing system using Halon or an inert gas. Where enclosures are not to be provided, equivalent arrangements will need to be put in place to cover the engine room as a whole.

Fire protection is, of course, a technically spe-cialised field. In addition, the detailed requirements for engine rooms will vary significantly from country to country depending on local codes of practice and legislation. Detailed guidance is, therefore, beyond the scope of this publication.

5.5 Installation Works – Civil

A number of items have been mentioned in section 4.9.4. These and further items are discussed in this section.

5.5.1 Foundations

Generating Sets The foundations for generating sets not only have to be adequate to support the weight of the plant but also have to have the strength to withstand the torsional forces that can arise under electrical fault conditions. Advice must be sought from the manufacturer of the chosen generating set.

Where IC engines are to be used, adequate vibration isolation arrangements will need to be incorporated into the fixing of the steel skid, on which the generating set is mounted, to the concrete base provided.

Waste Heat Boilers In a retro-fit project, it is not unusual to remove an existing conventional boiler and replace it with the waste heat boiler of the CHP plant. Care should, however, be taken when an existing base is to be re-used, as the weight of the waste heat boiler is likely to be considerably greater than the weight of the boiler it replaces.

5.5.2 Brick and Concrete Chimneys
The corrosive nature of the exhaust gases from engines has been discussed in some detail in section 5.2.3. It is, therefore, recommended that where existing plain brick or concrete chimneys are to carry engine exhaust gases, they are fitted with stainless steel flue liners.

5.5.3 Lifting Beams
The engines of CHP plants require regular stripping down and occasionally the removal of heavy components for maintenance purposes. It will, thus, be necessary to install a suitably sized lifting beam along with the appropriate tackle

to enable heavy components to be lifted up and away from the engine. Advice on the likely loadings and the lifting arrangements that are needed should be sought from the manufacturer of the chosen engine.

5.5.4 Noise Control

Large IC Engines Due to their relatively large physical size, it may not be practical to fit sound attenuating enclosures over large IC engines. Under these circumstances, the fabric of the engine hall will have to provide the required level of noise control. Designing and subsequently constructing a room or building that provides such a high level of sound attenuation will require that special attention is given to the choice of materials and the quality of the workmanship used in the construction works. In addition, particular care will need to be taken in the sealing of all holes through which services pass and the sealing of doors, which will probably need to be made of steel.

It is advisable, in this situation, to seek advice from an expert in noise control.

General Even when engines are fitted with enclosures, it may be unacceptable to have fixed openings for ventilation in the walls or doors of an engine room, at locations where there are non-industrial buildings close by. Where an engine is to be installed in an existing boiler room which is adjacent to residential accommodation, for example, it will be necessary to fill all fixed ventilation openings and install a mechanical ventilation system with in-duct sound attenuators to provide the necessary level of noise control.

5.6 Final Estimate of Financial Returns

5.6.1 Accuracy Required

The Calculation Approach In terms of investigation and design work, accuracy in estimation costs money. To a certain extent, therefore, the level of accuracy required in the prediction of energy savings/revenue depends on the magnitude of the associated capital spend. In addition, even utilising the most rigorous approach with the assistance of computer modelling, the accuracy that can be achieved in the prediction of energy savings/revenue over the lifetime of a CHP project is limited to perhaps ±20%.

This is, nevertheless, no reason to take a sloppy approach to the prediction work. In judging the financial success of the investment, most people will look at the savings achieved over the first few years of the operation of the scheme and will probably expect annual returns to be within 10% of the predictions made at the time of final scheme approval. Achieving this level of prediction accuracy with any energy efficiency investment is not easy. For a CHP scheme it is difficult. It is for this reason that such a rigorous approach to the calculation of energy savings/revenue has been set out in chapter 4.

When this approach is applied correctly, it should be possible to achieve a ±10% level of prediction accuracy for initial savings and revenue, where energy supply markets are stable and corrections are made for variations in demand from one year to the next.

Input Data No matter how sophisticated a calculation procedure is, it is obvious that the use of poor input data will produce poor predictions. At this final stage of scheme development it is, thus, vital that correct and up to date information is used for the exact make and model of each selected component of the CHP plant.

The greatest potential for error, however, lies in the prediction of future energy price inflation. The best advice on this issue, at the current time, is do not incorporate any judgements on relative energy price inflation but simply assume that energy prices will rise roughly in line with general price inflation.

Risk In comparison with most business investments, a combined heat and power scheme offers a low risk investment opportunity. There are, nevertheless, risks and these risks need to be quantified. For all but the smallest of schemes, therefore, a full risk analysis will be required. The issue of risk analysis and the calculations that are appropriate to CHP schemes are discussed in depth in insert panel 2 of this chapter. The importance of providing an analysis of risk as part of a cogeneration project proposal, cannot be over emphasised. Risk is a key issue for senior management and there is no better way to build the confidence of senior managers in a CHP scheme than through the presentation of a comprehensive risk analysis for the project.

5.6.2 Checklist

Listed below are a few reminders of the items to check for when compiling the final estimate of financial returns for a CHP project:

- For IC engines, make sure that the water temperature that can be generated from engine jacket cooling is adequate for the heating application served, all year round.
- Be confident that the appropriate number of bins has been selected for calculation.
- Check that the plant performance information used in the final set of calculations is for the exact make and model of each item of equipment selected.
- Check that all parasitic electrical losses have been allowed for, i.e. fuel gas compression and mechanical ventilation.
- Make sure that engine down time has been allowed for.
- Check that the guidance given in section 4.7.5 on maximum demand charge savings has be followed.
- Make sure that the energy price information used is based on figures quoted in writing by suppliers for the site in question, with the CHP scheme in operation.
- Be confident that plant size and equipment selection has been optimised through iteration.
- Make sure that maintenance cost estimates are based on firm quotations from the relevant suppliers.
- Check that adequate allowance has been made for the in-house staff time that will be required for the operation and management of the new installation.

5.7 Final Estimate of Capital Expenditure

5.7.1 Accuracy Required

Generally, senior management will be satisfied if capital costs at the completion of a project are within 10% of the estimate made at the time of final scheme approval. The accuracy that can be achieved for the estimate of overall implementation costs will, however, depend on the procurement route that it taken. The larger the proportion of the project that is tendered the greater the uncertainty regarding price.

The procurement options that are available are discussed at the beginning of the next chapter. What will already be obvious from the approach recommended for the selection of equipment and the prediction of financial returns, however, is that the components of a CHP scheme must be selected on a price/performance basis. It is likely, therefore, that the major items of plant for the scheme will be pre-selected on the basis of competitive quotations at the time of final design.

For this reason, provided that no item of the installation has been overlooked and provided that expert assistance has been sought for the evaluation of the necessary electrical reinforcement works, there should be no problem in achieving the ±10% accuracy required.

5.7.2 Checklist

Listed below are a few reminders of the items to check for when compiling the final estimate of capital expenditure for a CHP project:

- Get firm, written quotations for all the major components of the scheme from the selected manufacturers.
- Be confident that all plant selection and size optimisation work is now complete. Further work on optimisation should not need to be undertaken during the installation phase of the project.
- Make sure that estimates for the electrical installation works are based on design work undertaken by experts in the field. Check that these estimates include costs for any necessary reinforcement of the external supply network based on a firm quotation from the electricity utility.
- Ensure that all necessary items of ancillary plant have been allowed for.
- Check that all of the issues raised in sections 5.3, 5.4 and 5.5 of this chapter regarding interfacing and installation works are addressed by the final project design.
- Be confident that the installation works estimates are appropriate for the magnitude of the work to be undertaken.

5.8 Financial Appraisal

5.8.1 Appraisal Technique

Choice of Technique A detailed discussion of the various techniques available for the appraisal of the financial performance of an investment is beyond the scope of this book. A brief discussion of the main techniques in relation to the evaluation of CHP schemes will, however, be of value.

Simple Payback The evaluation of a scheme by considering the length of time needed to pay off capital investment through annual savings is still a widely used technique. The disadvantages of simple payback include:

a) Only constant annual savings/revenue can be analysed.
b) The effects of price inflation are ignored, though this is not necessarily of significance where inflation rates are only a few per cent per annum.
c) The savings/revenue produced by the scheme after payback of the investment are not considered.

As a crude way of reducing risk in the deployment of capital, however, many organisations still use simple payback as the overriding performance criteria for investment decisions. The argument is that, as the risk of misjudging financial returns increases the further into the future that predictions are made, a project with a fast payback is inherently less risky.

This logic is, unfortunately, flawed. An investment in a production line for a new product yet to be launched onto the market, may have a nominal payback of less than 24 months. The risks associated with such an investment are, however, potentially far greater than those associated with a CHP scheme that has a predicted payback of 5 years, for example.

CHP schemes, if evaluated and developed correctly, are low risk investments that offer substantial returns over a 10 to 15 year period. Repayment of capital, however, is not usually achieved in under 4 years. For these reasons, the simple payback investment criterion is highly unfavourable to CHP schemes.

Net Present Value Capital is a commodity like raw materials or manufacturing equipment and like any other commodity it costs money to obtain. For this reason, money owned now is of more value than money owned at some time in the future. In order to compare the worth of savings/revenue generated by a scheme in future years with the capital costs of a scheme now, the value of the future savings/revenue must, thus, be 'discounted'. The appropriate level of discount will depend upon the cost of servicing the capital utilised, the opportunity cost of not having the capital to use for other projects and the number of years into the future that the savings/revenue are to appear. The percentage figure used to discount the value of savings for each year that they occur into the future, is termed the 'discount rate'.

This discounting of the value of future savings/revenue is known as 'discounted cash flow (DCF)'.

Once a discounted cash flow analysis has been undertaken for an investment, the value of the savings/revenue over the lifetime of the installation can be totalled up and compared with the capital expenditure required for the scheme. This net savings minus cost figure is termed the 'net present value (NPV)' of the investment. In simple terms, a scheme with a positive NPV is worth financing whereas one with a negative NPV is not.

Judging the merit of competing schemes on the magnitude of the net present value they will produce is the most theoretically sound way of selecting which projects to invest in and which to reject. The financial characteristics of CHP schemes make NPV the most favourable investment appraisal technique to use for their evaluation.

Where capital resources are scarce, however, NPV may not be the appropriate appraisal technique, as it takes no account of the speed at which the return on an investment is to be achieved. In addition, it is important to realise that the accurate assessment of the appropriate discount rate to use, is vital if the NPV approach is to lead to the correct investment decisions being taken.

DCF Yield 'DCF yield' combines the technique of discounted cash flow with an indication of the rate at which return is being achieved on an

2

THE ANALYSIS OF RISK IN FINANCIAL PERFORMANCE

The first step in the execution of a risk analysis for a project is to determine which out of the parameters that have a major impact on the financial performance of the scheme are likely to change significantly over the investment period being considered. For a CHP project these parameters may include:

- Site electricity and heat demands.
- Energy purchase and export prices.
- Interest rates where non-fixed rate loans are used to finance the scheme.
- General price inflation.
- Capital cost overrun.
- Construction delays.

With the relevant parameters determined, step two is to decide a realistic maximum variation for each parameter. The exercise of this type of informed judgement is central to the analysis of any risk.

Whilst it is, of course, vital that variations for parameters are not seriously under-estimated, it is equally important that they are not completely over-estimated either, if the analysis is to be of any value. On issues such as interest rates and general price inflation expert advice should be sought.

With the extent of the possible variation in input data decided, the savings/revenue, operating cost and capital expenditure calculations can be repeated for the combinations of parameters that are likely to occur. Again,

judgement must be used to determine which sets of circumstances have a reasonable chance of occurring together. The simultaneous occurrence of the worst case for each individual parameter is extremely unlikely.

The requirement to undertake calculations for a number of different permutations, particularly the energy savings/revenue calculations, makes the execution of a comprehensive risk analysis impractical without the assistance of a computer programme for all but the simplest of CHP schemes. The application of computers to the evaluation of cogeneration schemes has been discussed in insert panel 3 of chapter 4.

With the calculations complete, a second evaluation of which combinations of circumstances are a realistic possibility can be undertaken. Then finally, the worst, best and most likely cases can be compiled for the financial performance of the project.

A summary only of the findings of the risk analysis is all that usually needs to be presented in the main report produced for a CHP scheme proposal. The detailed results should, however, be included within the appendices of the report in the form of a series of graphs that demonstrate the sensitivity of the financial performance of the scheme to variations in the studied parameters. The extent of the data contained in the appendices should be sufficient to demonstrate the comprehensive nature of the risk analysis that has been undertaken, thereby building overall confidence in the proposal.

investment. A number of DCF analyses are undertaken using different discount rates until, through iteration, the rate that produces a zero net present value for the scheme is determined. This discount rate is termed the DCF yield or 'true internal rate of return'. Naturally, the higher the DCF yield for a scheme the faster the return on investment.

Although savings/revenue over the lifetime of an installation are taken into consideration, low capital cost schemes with relatively short

lives but large annual savings/revenue will be favoured by this appraisal technique i.e. not CHP schemes. For this reason, CHP projects will not compete well against other projects when DCF yield is used as the investment criterion.

Where capital is not a scarce resource within an organisation or where schemes can be financed externally without diverting capital from other projects, then the NPV appraisal technique should always be utilised.

5.8.2 Risk

As has been mentioned in section 5.6, the proper evaluation of risk is crucial to persuading senior management to have confidence in a proposal. In addition, a clear demonstration that investment in combined heat and power, whilst having a relatively long payback, is of a low risk nature will play an important part in helping the scheme compete for capital funds. What needs to be done in the execution of a risk analysis for a CHP scheme is detailed in insert panel 2 of this chapter.

5.9 Environmental Impacts

5.9.1 Emissions

Concentration Limits for Noxious Emissions
Emissions from the engines of CHP plant have been discussed in chapter 3 and typical levels of noxious emissions in exhaust gases have been given in tables 3.1(c) and 3.2(c). These figures should be compared with the concentration limits for noxious emissions to meet emissions legislation expected to apply to generating plant by the year 2000, which are given in tables 5.9(a)

and 5.9(b) for the United Kingdom, Germany, Japan and California in the USA.

Interpretation of Noxious Emissions Data
Unfortunately, data on exhaust emissions are presented in a number of different forms, which can make comparisons between engines and checks that emissions will meet legislative requirements difficult.

Essentially emissions data can be presented in the following two forms:

a) Volume based i.e. parts per million volume by volume (ppm v/v) and mass per unit volume (g/m³)

Where this format is used, the temperature and pressure of the exhaust gases must be specified and whether the gas is with or without water vapour. The reference figures usually utilised are temperature 0°C, pressure 101.3 kPa and without water vapour. In addition, the oxygen content of the exhaust gases is also defined to prevent compliance with legislation being achieved by introducing fresh air into the exhaust stream simply to dilute the concentrations of noxious emissions. The reference figures most

Table 5.9 (a) *Emissions Limits expected to apply to IC Engine based Generating Plant by the Year 2000*

Item	Maximum Permitted Levels of Emissions[1]			
	United[2] Kingdom	Germany[3]	Japan[4]	California[5] USA
Nitrogen Oxides – ppm v/v	360	370	550	20–110
Carbon Monoxide – ppm v/v	350	190	570	60–250
Unburned Hydrocarbons – ppm v/v	180	80	250	130–320

Notes

[1] Emissions levels are for generating plant operating on natural gas. Levels expressed as ppm v/v are for dry gas conditions, adjusted to an exhaust gas oxygen content of 15%. For both ppm v/v and mg/nm³ gas temperature and pressure are taken at reference values of 0°C and 101.3 kPa respectively. Nitrogen oxides expressed as NO_2. Unburned hydrocarbons are non-methane.

[2] Based on UK Secretary of State's Guidance – Compression ignition engines, 20–50 MW net rated thermal input, PG1/5(91) February 1991.

[3] Erste Allgereine Verwaltungsvorschrift 2nm Blm SchG Technische Anleitung Luft, national minimum standard (for plant of <3MWt input the NO_x limit is 730). Proven advances in technology will, however, automatically lead to the reduction of these limits in future.

[4] MOC standards @ 1996.

[5] Limits apply to new plants and are assessed on a case by case basis.

Table 5.9 (b)   *Emissions Limits expected to apply to Gas Turbine based Generating Plant by the Year 2000*

Item	Maximum Permitted Levels of Emissions[1]			
	United[2] Kingdom	Germany[3]	Japan[4]	California[5] USA
Nitrogen Oxides – ppm v/v	60	75	85	3–5
Carbon Monoxide – ppm v/v	–	80	–	–
Unburned Hydrocarbons – ppm v/v	–	–	–	–

Notes

[1] Emissions levels are for generating plant operating on natural gas. Levels expressed in ppm v/v are for dry gas conditions, adjusted to an exhaust gas oxygen content of 15%. For both ppm v/v and mg/nm^3 gas temperature and pressure are taken at reference values of 0°C and 101.3 kPa respectively. Nitrogen oxides expressed as NO_2. Unburned hydrocarbons are non-methane.

[2] Based on UK Secretary of State's Guidance – Gas turbines, 20–50 MW net rated thermal input, PG1/4(91) February 1991.

[3] Technische Anleitung Luft 1994, national minimum standards for plant of <100 MWt input. Case by case determinations may result in the application of lower limits.

[4] More stringent limits apply in major cities and industrial areas. In Tokyo, for example, limits for NO_x range from 28 to 41 ppm v/v.

[5] Limits apply to new plants and are assessed on a case by case basis. For existing plants, limits range from 9 to 42 ppm v/v depending upon location.

often used are 5% oxygen for IC engines and 15% oxygen for gas turbines.

b) *Fuel or work based* i.e. mass per unit of work produced by the engine (g/kWh) and mass per unit of fuel energy consumed by the engine (g/GJ).

From the above, it will be seen that great care needs to be taken to ensure that the two sets of figures are truly comparable. Appendix 2 gives procedures for converting from one form of emissions data to another.

Carbon Dioxide Emissions   A switch to the use of cogeneration at a site will increase the local emissions of carbon dioxide (CO_2). When remote power station emissions are also taken into consideration, however, aggregate emissions can be dramatically reduced as a result of the more effective use of fossil fuels. Table 5.9(c) illustrates the benefits.

From the table it will be noted that CO_2 emissions for a gas burning CHP plant are less than 50% of those for the combination of remote coal

generated electricity and local gas generated heat. Nuclear generated electricity does, of course, produce zero CO_2 emissions although it does have other significant environmental problems associated with it.

It will also be seen from the notes that accompany table 5.9(c) that CO_2 emissions for natural gas are just 57% of those associated with coal. Hence, a major switch to any technology that burns natural gas will result in a dramatic reduction in CO_2 emissions.

Sulphur Dioxide and Nitrogen Oxides Emissions
The impact of cogeneration on noxious emissions is illustrated in tables 5.9(d) and 5.9(e) for sulphur dioxide (SO_2) and nitrogen oxides (NO_x) respectively.

From table 5.9(d) it will be noted that the combustion of natural gas produces virtually zero SO_2 emissions. A shift towards the greater use of natural gas, whether in central generation or in CHP will, therefore, dramatically reduce emissions of SO_2.

In the case of nitrogen oxides, a switch to gas

Table 5.9(c) *Impact of CHP on Carbon Dioxide Emissions*

Total emissions of carbon dioxide in kg associated with the use of 1,000 kWh of electricity and 2,000 kWh of heat at a site.

Heat Generating Plant and Fuel Type	CO₂ emmisions to produce 2,000 kWh of heat	Electricity Generating Plant and Fuel Type					
		Boiler & ST – Coal	Boiler & ST – RFO	CCGT– Gas	Boiler & ST – Nuclear	Gas Turbine CHP – RFO	Gas Turbine CHP – Gas
CO₂ emissions to produce 1,000 kWh of electricity		*1,070*	*890*	*450*	*0*	*–*	*–*
Boiler – RFO	*660*	1,730	1,550	1,110	660	–	–
Boiler – Gas	*460*	1,530	1,350	910	460	–	–
Condensing Boiler – Gas	*410*	1,480	1,300	860	410	–	–
Electrode Boiler – NE	*0*	1,070	890	450	0	–	–
Gas Tubine CHP – RFO	*–*	–	–	–	–	1,060	–
Gas Turbine CHP – Gas	*–*	–	–	–	–	–	730

Notes

1. Nominal electricity generating efficiencies and associated emissions per unit of fuel energy consumed, taken as:

Generating Plant	Fuel	Efficiency – %	Emissions – kg/GJ		
			CO_2	SO_2	NO_x
Boiler and ST	Coal	33	88.4	1.14	0.43
Boiler and ST	RFO	33	73.3	1.30	0.20
CCGT	Natural gas	45	50.6	0	0.10
Boiler and ST	Nuclear	–	0	0	0
Gas turbine CHP	RFO	25	73.3	1.30	0.15
Gas turbine CHP	Natural gas	25	50.6	0	0.10

2. Nominal heat generating efficiencies and associated emissions per unit of fuel energy consumed, taken as:

Generating Plant	Fuel	Efficiency – %	Emissions – kg/GJ		
			CO_2	SO_2	NO_x
Boiler	RFO	80	73.3	1.30	0.20
Boiler	Natural gas	80	50.6	0	0.05
Condensing boiler	Natural gas	90	50.6	0	0.05
Electrode boiler	NE	–	0	0	0
Gas turbine CHP	RFO	50	n/a	n/a	n/a
Gas turbine CHP	Natural gas	50	n/a	n/a	n/a

3. NO_x as NO_2 equivalent.
4. Nominal electrical distribution losses for remote generated electricity taken as 10% of distributed electricity.
5. Abbreviations: CCGT – Combined Cycle Gas Turbine
                   Gas   – Natural Gas
                   NE   – Electricity, Nuclear Generated
                   RFO   – Residual Fuel Oil
                   ST   – Steam Turbine

158

Table 5.9(d) *Impact of CHP on Sulphur Dioxide Emissions*

Total emissions of sulphur dioxide in kg associated with the use of 1,000 kWh of electricity and 2,000 kWh of heat at a site.

Heat Generating Plant and Fuel Type	SO₂ emissions to produce 2,000 kWh of heat	*Electricity Generating Plant and Fuel Type*					
		Boiler & ST – Coal	Boiler & ST – RFO	CCGT–Gas	Boiler & ST – Nuclear	Gas Turbine CHP – RFO	Gas Turbine CHP – Gas
		SO₂ emissions to produce 1,000 kWh of electricity					
		13.8	*15.8*	*0*	*0*	–	–
Boiler – RFO	*11.7*	25.5	27.5	11.7	11.7	–	–
Boiler – Gas	*0*	13.8	15.8	0	0	–	–
Condensing Boiler – Gas	*0*	13.8	15.8	0	0	–	–
Electrode Boiler – NE	*0*	13.8	15.8	0	0	–	–
Gas Tubine CHP – RFO	–	–	–	–	–	18.7	–
Gas Turbine CHP – Gas	–	–	–	–	–	–	0

Notes
See table 5.9(c).

Table 5.9(e) *Impact of CHP on Nitrogen Oxides Emissions*

Total emissions of nitrogen oxides in kg associated with the use of 1,000 kWh of electricity and 2,000 kWh of heat at a site.

Heat Generating Plant and Fuel Type	NO_x emmisions to produce 2,000 kWh of heat	Electricity Generating Plant and Fuel Type					
		Boiler & ST – Coal	Boiler & ST – RFO	CCGT–Gas	Boiler & ST – Nuclear	Gas Turbine CHP – RFO	Gas Turbine CHP – Gas
		NO_x emissions to produce 1,000 kWh of electricity					
		5.21	*2.42*	*0.89*	*0*	–	–
Boiler – RFO	*1.80*	7.01	4.22	2.69	1.80	–	–
Boiler – Gas	*0.47*	5.68	2.89	1.36	0.47	–	–
Condensing Boiler – Gas	*0.42*	5.63	2.84	1.31	0.42	–	–
Electrode Boiler – NE	*0*	5.21	2.42	0.89	0	–	–
Gas Tubine CHP – RFO	–	–	–	–	–	2.16	–
Gas Turbine CHP – Gas	–	–	–	–	–	–	1.44

Notes
See table 5.9(c).

fired CHP can, in fact, result in the generation of greater emissions than a number of other combinations of technologies. The nominal power generating efficiency figure of 25% chosen for the gas turbine CHP plant is, however, less than can be achieved by a large, well engineered system. Such a system, when gas fired, would produce NOx emissions roughly equal to those produced by the combination of gas fired combined cycle gas turbine and gas fired condensing boilers.

5.9.2 Noise
Noise generated by CHP plants and the measures that need to be taken to control that noise, have been discussed elsewhere in the relevant sections of the book. The issue of what noise levels will be acceptable outside an engine room in various situations has, however, not yet been covered.

Whether a certain level of noise is acceptable, of course, is a purely subjective question and the answer will vary from person to person. Experience in the US has, however, enabled a set of design criteria for noise to be compiled. Table 5.9(f) presents the data.

The figures given in table 5.9(f) are provided *solely for the purposes of provisional planning*. It is, of course, vital to determine the specific requirements that apply to a site by consulting the appropriate local regulatory body, before de-

tailed design work is undertaken.

The execution of a site noise survey prior to the implementation of a scheme can prove useful in discussions with regulatory bodies on the noise limits that are to apply. The results of such a survey may also be extremely valuable should complaints concerning noise arise once the CHP installation is in operation. It is not unknown for complaints of increased noise levels, following the installation of a new item of plant, to be received before the plant has been run for the first time!

Table 5.9 (f) also gives sound pressure levels in each octave band at which there is the risk of ear damage. Ear defenders must, of course, be obligatory in all engine rooms where sound pressure levels approach these values (this will be the case in all engine rooms where an acoustic enclosure has not been installed over the engine). It is, however, advisable to make the use of ear defenders mandatory in all engine rooms whether or not acoustic enclosures are in use. An exception to this rule can be made where smaller IC engine installations are concerned as the attenuation provided by the engine enclosure may reduce noise to totally safe levels.

5.9.3 Vibration
Once again, the vibration generated by CHP plants and the measures which need to be taken

Table 5.9(f)   *Design Criteria for Outdoor Noise in Various Situations*

Situation	Maximum Sound Pressure Levels for Acceptability – dB							
	37.5–75 Hz	75–150 Hz	150–300 Hz	300–600 Hz	600–1,200 Hz	1,200–2,400 Hz	2,400–4,800 Hz	4,800–10,000 Hz
Highly Critical Hospital or Residential Area	70	49	38	35	34	33	33	33
Residential Area at night	72	57	47	40	38	38	38	38
Residential Area during the day	75	62	52	45	43	43	43	43
Commercial Area	78	68	60	55	51	47	44	43
Industrial Area	85	82	76	72	70	68	66	66
Risk of Ear Damage	110	102	96	94	94	94	94	94

(Reproduced with permission from the American Society of Heating, Refrigerating, and Air Conditioning Engineers).

to control the transmission of that vibration, have been discussed elsewhere in the relevant sections of the book.

Annoyance from transmitted vibration arises from two sources. Firstly, from the physical sensing of the transmitted movement and secondly from local vibration induced noise. As the actual field measurement of vibration generated by plant is difficult, it is not usual to talk of vibration design criteria in terms of physical movement. The key is to ensure that the vibration isolation arrangements provided for the item of plant in question meet guidelines that have been produced as a result of experience.

A detailed discussion of the arrangements needed for the various components of a CHP system is beyond the scope of this book. Advice can, however, be sought from the manufacturers of equipment and also from the relevant ASHRAE Handbook (ref. 2).

5.10 Other Considerations

5.10.1 Management of the New Installation

In the case of small CHP plants, particularly when all operation and maintenance is contracted out, the management of the new installation will not be a major issue. On large plants, however, consideration will need to be given to the management that will be associated with the following items:

- Direction and control of site staff in the execution of operation and minor maintenance work.
- Direction and control of external maintenance contractors and negotiation of future maintenance contracts.
- Optimisation of plant performance, including decisions on variations to the overall control strategy.

The additional man time required to undertake the management of these items should be allowed for in the estimate of operating costs, at the appropriate level of seniority.

5.10.2 Operation and Maintenance

As has been emphasised throughout the book, the maintenance of engines and other items of CHP plant is a specialised activity which is un-

likely to be executed cost effectively by site staff. There will, however, be minor duties to be undertaken by site staff at larger CHP installations in connection with plant monitoring, taking oil samples, changing filters etc.

In the compilation of the estimate of operating costs for the scheme, the additional man time required for these duties must be allowed for, at an appropriately qualified plant technician's hourly rate.

5.10.3 Professional Fees

The professional fees associated with the investigation, design and project management of a CHP scheme may well amount to more than 10% of overall project costs. Even where a scheme is to be developed in-house, the required man hours at the appropriate hourly rates should be included within the overall capital cost estimate for the scheme.

Advice on the procurement of services from consultants is given in chapter 6.

5.11 Quality Assurance

5.11.1 The Importance of Quality Assurance

Project Development It is said that as a computational machine, the average human being is 99% accurate. As there are thousands of calculations to be undertaken in the development of a cogeneration scheme, it is virtually certain that errors in arithmetic, algebra and judgement will be made in the execution of this work. Some errors will be of little consequence others, however, may have a significant impact on the choice of technology, selection of equipment, prediction of savings or the estimation of capital costs. It is, therefore, of vital importance that *at all stages in the development of a scheme* adequate quality assurance procedures are in place.

Making the Case The importance to senior management of the appraisal of risk arising from factors that are external to a project has already been discussed earlier in this chapter. Of equal importance to senior management will, however, be the possibility that an error has been made in the development work undertaken on the project. The risks associated with these internal factors are just as real as those associated with external factors.

Evidence to demonstrate that a rigorous quality assurance system has been applied to the development work will play an important role in persuading senior management that they should have confidence in the CHP scheme proposal.

5.11.2 Achieving the Necessary Level of Quality Assurance

A detailed discussion of what comprises an adequate quality assurance system is beyond the scope of this book. The issues that are of particular importance to engineering schemes are, however, outlined below:

- Formal procedures, including the creation and signing of activity tracking sheets, are important. They are, however, of little value if not accompanied by the rigorous checking of:

  a) The input data used.
  b) Each individual calculation.
  c) The interpretation that has been made of the results.

- A different person from the project engineer must undertake all quality assurance work, including the checking of input data, calculations and interpretation of results.
- When computer programmes are used, care should be taken to ensure that the programme is being used in the correct fashion and that the application to be modelled is within the capability of the programme. In addition, approximate hand calculations should be executed to enable a final check of the reasonableness of computer predictions to be undertaken.
- Quality assurance should also encompass a review of the overall approach. This will require a step back from the detail to gain a perspective of the wider picture, to make sure that the fundamentals of a scheme are sound.

5.12 Preparing and Presenting the Proposal

5.12.1 The Importance of Proposal Presentation

The factors that weigh against a combined heat and power project when it comes to competing for capital resources within an organisation, have been discussed in the insert panel of chapter 2. To tilt the balance back in favour of a CHP scheme it is vital that first class presentational skills are deployed in support of the project proposal.

5.12.2 Form of Presentation

No form of presentation has the potential to be more persuasive than a person to person meeting. Small CHP projects may not be of sufficient magnitude to warrant a meeting with senior management. In the case of a major scheme, however, it is vital that a verbal presentation is made to the relevant business managers. A document alone, no matter how well written and produced, simply will not have the same impact.

5.12.3 Preparing the Proposal

Executive Summary In devising the content of the proposal to be presented, it is important to recognise that time is usually in short supply to senior management. Even for major schemes, therefore, the primary costs and benefits associated with the project should be presented in the form of a one to two page executive summary. This summary should be at the front of the overall proposal document or, better still, bound separately.

In a few minutes, through reading the executive summary, senior managers should get a clear understanding of:

- What, in general terms, is encompassed by the project.
- The capital expenditure required.
- The financial returns predicted.
- The financial performance of the investment.
- The risks associated with the project.
- Other non-financial but pertinent benefits that will be generated by the scheme.

In selecting the other benefits to be featured in the executive summary, close attention must, of course, be paid to the interests and responsibilities of the managers who are to receive the proposal.

Main Report The main report will probably only be read by the one or two managers in the management team who have a technical background. It should be remembered, however, that whilst these people may be fully qualified engineers, many of the issues and much of the technology relating to CHP may be new to them. It is important to keep this in mind when writing the more detailed parts of the report.

The issues that need to be covered in the main report will include:

- Choice of engine and ancillary plant technology.
- Sizing and selection of system components.
- Prediction of energy cost savings/revenue.
- Estimation of operating costs.
- Estimation of capital expenditure.
- Evaluation of risk.
- Financial performance of the investment.
- Environmental impacts.
- Management and staffing considerations.
- Quality assurance in the execution of the development work.

Appendices The appendices should always be bound in a separate volume to the main report, so that tables, charts and figures can be referred to without the need to page backwards and forwards through a single volume. The temptation to make the appendices voluminous should be resisted. The inclusion of all calculations undertaken will simply not be practical. As a guide the following might be included in the appendices:

- Electricity and heat demand profiles for the site, for typical days.
- A brief description of the approach taken to the calculation of energy cost savings/revenue predictions.
- Data sheets on the selected CHP system components.
- Current and expected fuel contract terms.
- The detailed results of the financial performance risk analysis.

Production Quality The quality with which the written proposal is produced and the clarity of the English used, will play an important role in determining the overall perception that senior management will form of the scheme. Never underestimate the importance of these items to gaining approval for a scheme.

5.12.4 Making the Presentation

Whether visual aids are required to augment the verbal presentation will be a matter for personal judgement. If it is considered that they are, they should be kept to a minimum. A flip chart or board to write on will almost certainly be of use in helping to amplify some of the issues being discussed.

In circumstances where someone other than you is to make the presentation, it is vital that they are not only fully briefed but also that they are convinced of the benefits of the scheme and are one hundred per cent behind it.

In preparing for the presentation, check that:

- A comprehensive written proposal to support the verbal presentation has been produced and distributed to senior management well in advance of the meeting.
- The content of the presentation places emphasis on the issues of particular interest to those attending the meeting.
- Visual aids are produced and carefully checked well in advance of the presentation.
- Sufficient practice of presentation delivery has been undertaken.

5.12.5 Building a Consensus for the Scheme

A detailed discussion of the psychology that is involved in building a consensus for a cogeneration scheme is, mercifully, beyond the scope of the book. The following items may, however, be worth considering:

- The interest and support of key senior managers should be sought early in the development of a scheme and well in advance of any formal project presentation.
- The benefits that will be generated by the scheme in each individual area of the organisation's activities should be identified and discussed with the personnel concerned.
- In the unusual situation where a CHP scheme may have a negative impact on an activity, the potential difficulties should be squarely addressed and measures to mitigate the impact determined and discussed with those concerned.
- Steps should be taken to raise the profile of energy efficiency and the environment in general within the organisation, to build support for the concept of combined heat and power.

- Combined heat and power projects undertaken by similar organisations or, better still, competitors should be brought to the attention of senior management.

6

SYSTEM PROCUREMENT

Component selection and design optimisation are of vital importance to the success of a CHP scheme. Of equal importance, however, is the approach that is taken to the implementation of the project.

SYSTEM PROCUREMENT

6.1 Procurement Options

6.1.1 Financing Alternatives

The procurement options that are available will, to a large extent, depend on the method to be used for financing the scheme. The alternative financing methods for CHP projects are discussed below.

a) Internal Funding Competing for and securing the necessary finance from the general capital funds available within the purchasing organisation is the most common method used for the funding of cogeneration schemes.

Advantages
- 100% of the financial returns are retained by the purchaser.
- The purchaser has complete choice in the selection of equipment.
- The purchaser has complete freedom in the choice of procurement option.
- Minimum cost of capital.

Disadvantages
- 100% of the risks associated with the scheme are borne by the purchaser.
- Competition for general capital funds will usually be fierce, which may lead to project delays or the rejection of the CHP scheme.

b) Hire Purchase of Plant The suppliers of packaged CHP plants or generating sets will often be able to organise third party financing for the major items of plant. Alternatively, a commercial bank can be approached directly to provide the necessary funds.

Advantages
- 100% of the financial returns are retained by the purchaser.
- Third party finance can usually be secured quickly and with little difficulty, thereby minimising delays in project implementation.
- The purchaser has complete choice in the selection of equipment.

Disadvantages
- 100% of the risks associated with the scheme are borne by the purchaser.
- Where the finance is organised through the suppliers of plant, there may be limitations placed on the procurement options open to the purchaser.
- Commercial finance charges will be paid.

c) Lease of Plant Some suppliers of packaged CHP plants or generating sets will be prepared to lease the plant to a purchaser. At the end of the lease period, which might be 5 or more likely 10 years, the purchaser will have the option to purchase the plant, extend the lease or terminate the agreement.

Advantages
- 100% of the financial returns are retained by the purchaser.
- Leases can usually be secured quickly and with little difficulty, thereby minimising delays in project implementation.

Disadvantages
- 100% of the risks associated with the scheme are borne by the purchaser.
- The selection of equipment will be limited to those suppliers prepared to offer leases.
- Commercial finance charges will be paid.

d) Shared Savings Some suppliers of packaged CHP plant or, alternatively, energy supply companies will be prepared to take responsibility for part or all of the necessary project finance in return for a share of the financial returns generated by the scheme. There are many variations on this type of arrangement, ranging from the provision of 100% of the finance by a packaged CHP plant supplier to the formation of a joint venture company with the local electricity or gas utility.

The percentage of the financial returns retained by the purchaser will depend on the proportion of capital financed by each party. In arrangements where the other party takes 100% of the risks, the percentage of the financial returns retained by the purchaser can be as low as 20%.

Advantages
- Some or all of the risks associated with the scheme are transferred elsewhere.
- Shared savings contracts can usually be put in place quickly and with little difficulty, thereby minimising delays in project implementation.

Disadvantages
- Only part of the financial returns are retained by the purchaser.
- In the case of an agreement with a supplier of packaged CHP plant, the selection of equipment will be limited to those suppliers prepared to enter into shared savings contracts.
- The purchaser may be required to procure the project under a turnkey design and build contract with the other party.

6.1.2 Optimum Financing Choice

The fundamental choice that has to be made is between alternatives a), b) and c) of section 6.1.1, where 100% of the risks of the scheme are borne by the purchaser, and alternative d) where the risks are shared.

Risk sharing is equivalent to taking out insurance cover. Organisations take out insurance when the outcome of a particular set of events would result in a financial loss that could not comfortably be borne by that organisation. So, for example, large corporations may not insure their motor vehicles (other than to meet basic legal requirements) but would, almost certainly, insure their headquarters building.

The key in deciding whether to share risk is to ask whether the maximum likely potential loss associated with the project could comfortably be borne by the purchasing organisation. In the case of a CHP scheme the maximum likely potential loss relates to the worst case prediction for savings/revenue. For all but the largest of schemes, where energy export is to form a major component of savings/revenue, this maximum likely potential loss can usually be borne by the organisation concerned with ease. There does not, therefore, seem to be any logical reason for electing to share risk and hence financial returns on CHP schemes.

The promoters of CHP schemes have, however, to live in the real world. In the real world, senior management are particularly concerned by risk and rightly so as the failure of a large project can seriously damage an individual's career prospects.

For this reason, the fact that the maximum likely potential loss associated with a CHP project could comfortably be borne by an organisation is irrelevant. As far as senior management are concerned, the key question is whether the benefits provided by the scheme outweigh the personal risks involved.

With this simple analysis of psychology in mind, the real advantage of shared savings schemes for the financing of CHP projects becomes apparent. To convince senior management to sanction a project where 100% of the risks are borne in-house, requires excellent persuasive and presentational skills supported by a thorough and reliable feasibility study. In contrast, where 100% of the risks are born by others, getting approval for a project is usually a formality.

In conclusion, therefore, shared savings agreements provide an easy and useful route to financing CHP schemes and to gaining the approval of senior management. Such agreements should, however, only ever be used when senior management cannot be persuaded to retain risk in-house and hence keep 100% of the cost savings generated by the scheme.

Where senior management can be persuaded to retain risk, the final choice between alternatives a), b) and c) will be determined by the cash position of the organisation and the comparative tax advantages of hire purchase versus leasing.

6.1.3 Procurement Methods

a) Conventional Project Procurement In the construction industry, the conventional method of procuring a construction project is for the purchasing organisation to take responsibility for the investigation, design and project management needed for scheme implementation. The construction works are undertaken by a contractor, selected on the basis of a competitive tender. An external consultant may or may not be used for the investigation, design and project management activities.

In its pure form, this method of procurement is not particularly well suited to the purchase of a retro-fit CHP system. The reason is that in the optimisation of component sizing and selection, firm decisions on the make and model of the major items of plant will already have been made before a tender for the complete installation can be undertaken. Is it not, therefore, possible to invite CHP plant or generating set suppliers to tender competitively for the complete installation. Conversely, normal mechanical contractors will not be particularly interested in tendering and taking responsibility for the complete installation, when such a large proportion of it is related to one preselected item, the CHP package or generating set. The two possible solutions to this problem are:

i) Purchase the CHP plant and the installation works under separate contracts.
ii) Negotiate a contract for the complete installation with the selected CHP package or generating set supplier.

Solution i) is a non-starter for anything but the smallest of projects, due to the potential for disputes should a scheme fail to perform as expected. Solution ii), negotiation, is thus the only practical option. For many organisations, however, the existence of compulsory competitive tendering rules for projects over a certain size will preclude the use of negotiation/single tender action, in all but the most exceptional of circumstances.

Where the CHP scheme forms part of a major new build or refurbishment project, the problems with tendering do not arise as the supply and installation of the necessary plant can be incorporated into the tender for the overall works, in the normal fashion.

The pros and cons of the conventional project procurement approach when used for a CHP scheme are as follows:

Advantages
- The purchaser, either directly or via a consultant, retains control over the sizing and selection of the major system components.
- The design of the project is essentially complete before construction commences. All potential problems should, therefore, have been identified in advance and the necessary steps taken to mitigate these problems.
- Responsibility for project development rests with one party, the purchaser (or his consultant). There is, therefore, no potential for confusion over who is tasked with design optimisation.
- The party designated to undertake project development, the purchaser, has an interest in the financial returns generated by the scheme and hence will be motivated to achieve the optimum design.

- As a result of these advantages, this procurement method has a high likelihood of producing a project that makes the most of the potential for savings/revenue that exists at a site.

Disadvantages
- Responsibility for the project is split between design and installation, leading to the potential difficulty of determining fault should the system fail to perform as expected.
- Competitive tendering for the installation works associated with the CHP scheme may not be practical on retro-fit projects.
- Implementation may take longer due to the separate design and construction phases of the project.
- A significant sum of money will be spent by the purchaser on project development, whether or not outside consultants are used.

b) Supply and Install In manufacturing industry, it is usual in the procurement of a major item of plant to select a supplier at an early stage in project development and then work closely with the supplier in the design of the project. The chosen supplier then supplies and installs the necessary equipment. The employment of outside consultants is usually limited to areas of the project where some specialist assistance is needed.

From the very nature of supply and install, it will be obvious that a competitive tender for the project cannot be undertaken. Negotiation, therefore, has to be used.

Advantages
- The purchaser retains control over the sizing and selection of the major system components.
- The design of the project is essentially complete before construction commences. All potential problems should, therefore, have been identified in advance and the necessary steps taken to mitigate these problems.
- Less money will be spent by the purchaser on project development.

Disadvantages
- Responsibility for the design of the project is split, leading to the potential difficulty of determining fault, should the system fail to per-

form as expected.
- No one is formally designated to undertaken project design and development, making the full optimisation of the scheme less likely.
- Competitive tendering for the installation works associated with the CHP scheme is not possible.
- Implementation may take longer due to the separate design and construction phases of the project.

c) Design and Build A procurement route becoming ever more popular in the construction industry is 'design and build'. The main feature of this method is that the design and construction phases of the project are bought together into a single phase, with responsibility for design being transferred mainly to the contractor. In project development terms, the input from the purchaser is limited to the production of a scheme design and the writing of a performance specification. A competitive tender for project design and construction is then undertaken on the basis of the specification.

When applied to a CHP project, the companies invited to tender for the works will either be CHP/generating set packagers or, in the case of large schemes, major construction companies. With responsibility for detailed design being passed to the contractor, it is not possible for the purchaser to select the exact make and model for the major components of the scheme.

Advantages
- Responsibility for both the design and installation of the project rests with one party.
- Implementation time will be minimised due to the compression of design and installation into one project phase.
- Direct project development costs to the client will be minimised, though payment will still have to be made to the contractor for the design work that he undertakes.
- Competitive tendering for the entire installation can be undertaken.

Disadvantages
- The purchaser no longer has control over the sizing and selection of the major system components.
- The design for a project is rarely complete before construction commences. It is not, therefore, possible to identify all potential problems in advance and the risk of major difficulties arising on site is significantly increased.
- The party formally designated to undertake project development has no interest in the financial returns generated by the scheme (unless shared savings financing is being used) and hence will not be motivated to achieve the optimum design.
- As a result of these disadvantages, this procurement method has a low likelihood of producing a project that makes the most of the potential for savings/revenue that exists at a site.

6.1.4 Selection of Procurement Method

There is a common theme running through this book regarding optimisation of design through detailed evaluation and rigorous sizing and selection of equipment. The results that are achieved by a combined heat and power installation, perhaps more than any other engineering services project, depend on the effort and care that has been put into the development of the scheme. Experience has shown that additional investment in the development and design of a cogeneration project pays off handsomely once the plant is operational and producing financial returns.

It is vitally important, therefore, that one party is charged with responsibility for design optimisation and that the necessary time and resources are made available for the project development work. For this reason, where small schemes are concerned, 'conventional project procurement' should be preferred to 'supply and install', in spite of the initial cost savings that may be achieved using the latter procurement method.

In contrast, for large installations, the contractual advantages of 'design and build' make it the preferred option. To ensure that development work and, in particular, design optimisation is undertaken correctly it will be necessary to detail the actual methods to be used for development and design in the performance specification, in addition to the required end result.

As large installations tend to be technically complex, it is normal practice for the purchaser to hire an external consultant to produce the performance specification and to compile the necessary contract documentation. It is then usual for the consultant to be retained to monitor the performance of the successful contractor in the execution of the project development work, to ensure that the final design produced has been optimised for the site in question.

6.2 Preliminary Evaluation

The execution of a CHP feasibility study will take a considerable length of time and will cost a significant sum of money, regardless of whether it is executed in-house or by external consultants. It is, therefore, important that the potential for combined heat and power at a site is tested before a detailed feasibility study is commenced.

The preliminary evaluation can usually be undertaken by in-house staff with the assistance of the figures and accompanying tables held at the back of chapter 2. Alternatively, but less satisfactorily, a supplier of CHP plant can be invited to provide an opinion free of charge.

Once the potential for savings has been confirmed, then funds or in-house time can be allocated for the execution of the feasibility study.

6.3 The Feasibility Study

6.3.1 What is required
The purpose of a combined heat and power feasibility study is to test the technical and financial feasibility of installing a cogeneration system at a site. The data that needs to be collected, the calculations that must be undertaken and the cost estimates and savings predictions that have to be produced are all detailed in chapter 4.

6.3.2 Who should do it
Armed with this book and other publications on the technology, a professional engineer experienced in energy efficiency investigations and with a good knowledge of mechanical engineering and services project work will be able to undertake the CHP feasibility study. Unless he also has expertise in high voltage electrical switching and protection, however, outside assistance will be required in this area.

Even a small cogeneration project will be rel-

atively complex and will demand a significant level of capital expenditure. For these reasons, if you don't have the experience and knowledge outlined above you would be well advised to seek assistance. From reading this book, however, you will be in an excellent position to judge the quality and veracity of the advice you receive.

6.3.3 Preparing a Brief and Selecting an External Consultant

The Brief In preparing the brief for the feasibility study, resist the temptation to specify in exact detail every measurement and calculation that should be undertaken by the consultant. The brief should take the form of an invitation to produce proposals for the study. It should provide an outline of the duties you expect to be undertaken, which can be derived from the tasks discussed in chapter 4.

Short-listing the Consultants A short list of suitably qualified and experienced consulting engineering practices should be prepared based on:

a) Outline information on similar past study commissions undertaken.
b) The CV's of the engineering staff that would actually undertake the commission.
c) References taken by telephone, or preferably face to face, from past clients.

As the selection procedure will involve the input of a considerable amount of time, both from the engineering practices and from the purchaser, even for large projects the number of companies short-listed should be kept to perhaps no more than four.

The Presentation and Interview Based on the brief, the short-listed practices should then be invited to prepare and submit a detailed written proposal for the commission, to be followed by a verbal presentation and interview. The member of staff who would actually be allocated by the practice as lead engineer for the commission will be required to make the presentation.

The advantage of keeping the brief short and

relatively general will now become apparent. Based on a detailed appraisal of the proposal in relation to the guidance and advice given in chapter 4, it will be possible to form a relatively accurate opinion of the knowledge and expertise held by each company in the field of CHP and, in particular, CHP feasibility studies. The verbal presentation and interview can then be used to explore any areas of uncertainty and to get a first hand view of the engineer who would lead the engineering team used for the commission.

One important area to explore in the interview is the approach that the consultant intends to take to the prediction of energy cost savings/revenue. Remember, in all but the most straightforward of circumstances, the assistance of a suitable computer programme will be required in the calculation of savings, particularly if a risk analysis for the investment is to be undertaken.

Selection Criteria You will no doubt notice that, so far, there has been a conspicuous omission of one important item in the discussion on the choice of practice, namely price. Being a consultant myself, you might expect me to be in favour of selecting the practice first and then negotiating commission fees. Such an approach, however, is unlikely to secure the best value for money for the commission and, in any case, is unlikely to satisfy the selection requirements of most organisations. There is, in fact, nothing wrong with inviting practices to make sealed bids for an item of work, providing the purchaser is then in a position to select on the basis of value for money, rather than lowest price.

As part of the written proposal the consultancies will, thus, be required to provide a fixed lump sum price for the commission, inclusive of all expenses. In support of the price, each submission will also provide a breakdown, to the nearest hour, of the manpower resources of each grade of engineer allocated to each individual element of the commission. Finally, the hourly charge rates and any expenses that apply to the proposal will be stated.

Armed with the information provided in the written proposals, supplemented by the judgement formed from the presentation and interview, the purchaser will be in an excellent position not only to make the selection decision on the basis of competence and value for money but also to justify that decision to others.

The precise duties proposed by the consultant in response to the brief may not exactly meet the requirements of the purchaser. The detailed breakdown of man hours and statement of hourly charge rates submitted with the proposal will, however, provide a sound basis on which to agree the price implications of any adjustments to the engineering duties.

Cost In deciding what is a reasonable sum to pay for a detailed CHP feasibility study, the importance of getting the study right should always be held in mind. Whilst it is difficult to generalise, a study that comprehensively covers the issues raised in chapter 4 will probably cost in the region of 2 to 4% of the capital cost of the project.

This is a relatively small price to pay to ensure that the fundamentals upon which the whole scheme is to be developed are right.

6.4 Design and Specification of the Project

6.4.1 What is required
The level of design to be undertaken by the purchaser for the various procurement options is briefly discussed below.

Conventional Project Procurement With this procurement option the sizing and selection of all items of equipment is undertaken by the purchaser or his consultant. The major items will include:

- Packaged CHP plant or generating set.
- Heat recovery equipment.
- Noise control equipment.
- Fuel supply and treatment equipment.
- Gas compression plant, where required.
- Electrical switchgear and protection equipment.

The sizing and routing of the mechanical services required to interface the CHP system with conventional heating plant, is usually also undertaken by the purchaser. Responsibility for this detailed mechanical design work can, however, be passed to the contractor if preferred. In the case of the design of the necessary electrical services and any reinforcement work made necessary by the CHP installation, this has to be undertaken by the purchaser, or more likely his specialist consultant, as the sizing of electrical switchgear and protection equipment is dependent upon it.

The calculations that will need to be undertaken, the issues to be addressed and the decisions to be made are covered in chapter 5.

It is recommended that use is made of industry recognised standard specifications to detail the general standards that will have to be met by the installation in the areas of mechanical, electrical and civil engineering. This will leave the particular specification prepared for the contract free to concentrate on the issues that are pertinent to the installation of a cogeneration system.

In addition to the specification of plant and equipment, particular attention should be paid to the following items:

- Materials.
- Installation techniques and standards.
- The cleaning and flushing of existing hydraulic circuits.
- The requirements regarding disruption of site services and the need for out of hours working.
- The automatic control and monitoring facilities to be provided.
- Commissioning and testing of the hydraulic circuits.
- Commissioning and testing of the electrical switchgear and protection.
- Commissioning and testing of the CHP plant and automatic controls.
- Availability and plant performance guarantees to be provided by equipment suppliers.
- The terms and conditions of the maintenance agreement to be provided for the CHP plant.
- Sections 5.3, 5.4 and 5.5 of chapter 5 provide further guidance on the key issues that should be addressed.

In addition to the specification, drawings will, of course, be required to convey the design intent to the contractor. Typically, the drawings required will include:

- Mechanical and electrical schematics to illustrate the general arrangement and intercon-

nection of services.
- Mechanical, electrical and builders work layout plans to detail the positioning of plant and equipment, the sizing of ducts, pipes and cables and the approximate routing of services.
- Architectural plans, elevations and construction detail drawings where a new building is to be provided.
- Services co-ordination drawings, where necessary.
- Electrical line diagrams.

Supply and Install With this approach to procurement, the question of who is responsible for what is not clearly defined as the project tends to be developed through an informal arrangement with a supplier. The chances of important issues being overlooked at the design stage are, thus, greatly increased when using this procurement method. It is, therefore, important that the arrangements with the supplier are formalised in writing, with the responsibilities with regard to design being clearly set out.

The actual items of design work that must be undertaken by one or other of the parties are as detailed for 'conventional project procurement'.

There is another area where the supply and install approach can run into problems. Using this method it would be unusual for either the purchaser or the supplier to produce a detailed specification and accompanying drawings for the works before a contract is placed with the supplier. The risks associated with the purchase of a complex installation in the absence of a detailed specification, in terms of future contractual disagreements, is self evident.

To reduce those risks to an acceptable level, the supplier should be required to provide a detailed quotation for the works with accompanying schematic drawings and rough layout plans. Once agreed, these documents can then be used as the technical basis for the contract. The supplier's quotation should also refer to industry recognised standard specifications to detail the general standards that will be met by the installation in the areas of mechanical, electrical and civil engineering.

Design and Build In design and build, the contractor undertakes the bulk of the design work. The specification for the design and build approach is, therefore, concerned with setting out the performance that is to be achieved by a project and with specifying general installation standards. As has already been mentioned in section 6.1.4 however, in addition to the installation works, the specification for a design and build project also has to cover the approach to be taken to scheme development and, in particular, to design optimisation. The necessary duties are briefly discussed in section 6.4.3.

In the case of design and build, the drawings produced by the purchaser will be limited to perhaps mechanical and electrical services schematic drawings and rough layout plans.

6.4.2 Who should do it

The requirements in terms of qualifications, experience and knowledge are the same as those stated in section 6.3.2. Where a purchaser has no suitable personnel available in-house to tackle the design work, then he has two options. One is to hire an outside consultant. The other is take the design and build procurement route.

Even when a design and build procurement route is chosen, however, a professional engineer, experienced in engineering project work, will still be needed to devise the performance specification and to appraise the technical content of the tender returns provided by contractors.

6.4.3 Preparing a Brief and Selecting an External Consultant

The procedures for preparing a brief and selecting an external consultant to undertake the design of a CHP scheme will be similar to those used for the feasibility study. Reference should, therefore, be made to section 6.3.3. The particular issues that will need to be addressed for the design work are discussed below.

The Brief   Use should be made of an industry recognised form of agreement for the provision of consultancy services in relation to engineering design. In addition to this standard document, a particular brief should be prepared to outline the approach that is to be taken to the sizing of CHP plant, selection of equipment, optimisation of design etc. The list of necessary duties can be derived from the tasks discussed in chapter 5.

Short-listing of Consultants   The procedure will be exactly the same as used for the feasibility study, except that outline information on similar past design commissions will need to be appraised.

From the perspective of cost efficiency, however, where a practice has already been used for the feasibility study it makes sense to invite that company to undertake the design work. It is important, therefore, when selecting a practice for the feasibility study that only consultancies with the requisite design and project engineering capabilities are considered.

The Presentation and Interview   Again, the procedure will be similar to that used for the feasibility study. For the design commission it will be important to explore the following particular areas in the interview:

- The approach the consultant intends to take to design optimisation and, in particular, the relevant computer software at his disposal.
- The level of expertise held in electrical switching and protection in relation to the generation of electricity, particularly where high voltage works are involved.
- The consultant's knowledge of the items that need to be included in a particular specification for a CHP installation, as outlined in section 6.4.1.
- The consultant's knowledge and practical understanding of the commissioning and testing work that will be required, as outlined in chapter 7.

Selection Criteria   The criteria are exactly as those applied for the selection of a consultant to execute the feasibility study.

Cost   Where the consultant has undertaken the feasibility study, it may be possible to agree a fixed lump sum price for the design commission. Though unusual in some countries, this method of charging for design work has a fundamental advantage over the scale fee percentage system in that the consultant has no incentive to increase the capital cost of the project. Equally, the consultant has no disincentive to optimise the design and hence cut capital costs.

The design of a CHP installation is unusual for two reasons. Firstly, the work required to size and select equipment and then optimise those selections is particularly involved and time consuming, even with the assistance of a computer programme. On the other hand, project costs will be dominated by the supply of a few large items of plant making the amount of work on services sizing, routing etc. small in relation to overall capital costs. In assessing the level of fee appropriate for the work, therefore, consideration needs to be given to these counterbalancing features.

Where the consultant has undertaken the feasibility study and is then employed to produce a detailed design for the installation along with drawings and specifications, commission fees in the region of 4 to 6% of the capital cost of the project should be expected. Again, it needs to be emphasised that good design and full system optimisation will have a major impact on the financial performance that is eventually achieved by a completed CHP installation.

6.5 Tender Action

6.5.1 Single Tender
The options for procuring a CHP system have been discussed in section 6.1. When either option a) 'conventional project procurement' or option b) 'supply and install' are selected, there will be no practical alternative to single tender action as the choice of CHP packager/generating set supplier will have been taken at a relatively early stage in project development. Section 5.1 of chapter 5 provides guidance on the selection of CHP plant packagers and generating set suppliers.

The single tender action and subsequent contract for the installation works will be based on the specifications and drawings produced by the purchaser for the project or, in the case of supply and install, on a detailed quotation and scheme drawings prepared by the supplier.

6.5.2 Competitive Tender
In the case of procurement option c) 'design and build', a competitive tender will be conducted for the project. At the outset of tender action the principle that final selection will be based on value for money rather than lowest price must be firmly established within the purchasing organisation. No matter how carefully the tenderers are selected, the value for money provided by each offer will vary markedly from tenderer to tenderer, with the lowest price not necessarily providing the best value for money. It should be emphasised that, unlike a conventional installation contract, a design and build contract includes the detailed design of the installation. So if the design and plant optimisation work is poor the performance of the completed CHP installation will be poor.

CHP packagers or generating set suppliers should be short-listed for the tender, on the basis of the guidance given in section 5.1.

The process of tender evaluation and selection will be very similar to that used for the selection of consultants and should, therefore, include a presentation and interview for each tender. Again, in common with the selection of a consultant, the presentation should be made by the member of staff who would manage the project for the tenderer, if successful.

The tender returns provided should include a detailed breakdown of costs both for the design and installation elements of the project. An appraisal of these breakdowns will reveal whether or not the tenderer has placed the appropriate emphasis on each element of the contract and has a complete understanding of the requirements of the performance specification.

Armed with the information provided by the tender returns, supplemented by the judgement formed during the presentation and interview, the purchaser will be in a good position to chose the tenderer most likely to provide an installation that makes the most of the potential for savings/revenue that exists at the site. He will also be in an excellent position to justify that choice, should the chosen tender not be the lowest priced tender received.

6.6 Project Management

6.6.1 What is required
The purpose of project management is to get a project installed and performing in accordance with the specification, to time and to cost, with the minimum of disruption to existing site services. Much of what is required for the management of a CHP installation is, of course, the same as is required for any major engineering project. In the case of a cogeneration system,

however, the correct execution of commission and testing is of paramount importance. The whole of chapter 7 of this book is, therefore, dedicated to this single issue.

The role and responsibilities of the purchaser in project management will differ depending upon the procurement option used. These differing roles and responsibilities are briefly discussed below.

Conventional Project Procurement   Under this procurement option, the purchaser or his consultant will be required to undertake the classic role of project manager. This role includes responsibility for the following items:

- Overall construction control, to ensure that the installation works are undertaken in accordance with an agreed programme and that close liaison is maintained between the operators of the site and the contractor.
- Site supervision, to ensure that the installation conforms with the specification and drawings and to make sure that all specified tests are undertaken and that the test results are satisfactory.
- Financial management of the contract, sometimes with the assistance of a separate quantity surveyor.
- Approval of the final installation for handover to the purchaser.

Supply and Install   When this procurement option is selected the role of project manager is less formally defined. The role and responsibilities of the purchaser's project manager are, nevertheless, broadly similar to those defined above for 'conventional project procurement'.

Design and Build   For design and build projects much of the construction control and site liaison work will be undertaken by the contractor's project manager. In this case, the purchaser will provide a supervising engineer who will have responsibility for the following items:

- Supervision of project development to ensure that the work, particularly design optimisation, is undertaken in accordance with the performance specification.
- Witnessing of commissioning and testing to make sure that the tests are undertaken and

that the test results are satisfactory.
- Approval of claims for payment and the negotiation of admissible variations, sometimes with the assistance of a separate quantity surveyor.
- Approval of the final installation for handover to the purchaser.

6.6.2 Who should do it
Once again, the requirements are similar to those set out in section 6.3.2. Where procurement option a) or b) is used, however, in addition to the requisite technical knowledge and experience the manager of the project will need to be experienced in taking the lead role in the supervision of construction projects.

Whichever procurement route is to be used, a suitably qualified project manager must be appointed to act on behalf of the purchaser. Where in-house personnel are not available, an external consultant must be hired.

6.6.3 Preparing a Brief and Selecting an External Consultant

The Brief   For procurement options a) and b) it is likely that the same practice will undertake both the design and project management work. The selection of a consultant for both sets of duties should, thus, be made at the time of the appointment for the design commission. Equally, in the case of procurement option c), the selection of a consultant to prepare the performance specification and then provide engineering supervision of the contractor should be undertaken at the same time.

It will, therefore, be necessary to include the additional supervisory and management duties within the brief discussed earlier in section 6.4.3. For a CHP project, the particular brief prepared for the commission should emphasise the following items:

- For design and build projects, exactly what is expected of the consultant with regard to the supervision of project development work.
- For 'conventional' and 'supply and install' projects, the exact split of site supervision duties between the consultant and the clerk of works, where the clerk of works is provided by the purchaser.
- For all projects, the extent of the witnessing

of commissioning and testing that is to be undertaken by the consultant.

Cost Again, it may be possible to agree a fixed lump sum cost for the necessary project management or engineering supervision commission, where the consultant has undertaken the feasibility study. Otherwise, a scale fee percentage will need to be agreed. A CHP scheme will require an unusually high level of supervision during the installation phase of the project to ensure that disruption to existing services is minimised, that the specified commissioning and testing procedures are executed correctly and that the overall plant achieves the expected level of financial returns. Where full project management is required, it would not be unusual to pay from 2 to 4% of the capital cost of the installation, excluding the provision of a clerk of works.

7

COMMISSIONING
and
TESTING

Commissioning and testing are, of course, of great importance to the success of any engineering installation. For a CHP scheme they are simply fundamental.

COMMISSIONING AND TESTING

7.1 General

7.1.1 Requirements

A CHP installation is comprised of a number of relatively complex items of equipment which must, of course, be commissioned and tested individually. The success of the project including the achievement of the predicted financial returns will, however, be determined by the way these individual items of equipment perform together as a single CHP plant.

Such interdependence of equipment makes the commissioning of a CHP installation similar to the commissioning of an industrial process plant in terms of the length of time and level of sophistication required.

7.1.2 Roles and Responsibilities

Roles   Where required, the functional testing of individual components of the CHP system is undertaken by the relevant manufacturer at his works. The on-site commissioning and testing of the major items of plant is also undertaken by the relevant manufacturer but with the contractor for the project in attendance. The commissioning and testing of the installation works is, of course, undertaken by the contractor for the project.

Finally, overall plant performance testing and optimisation is undertaken by the contractor with the assistance of manufacturers as necessary.

Certification of the electrical switching and protection arrangements may be required before the installation can be connected to an external supply network. Where this is the case, the relevant electrical installation tests will be witnessed by a representative of the external supply utility.

Responsibilities   The contractor for the project is, generally, charged with responsibility for ensuring that all tests are executed and the results correctly recorded, whether the tests are carried out on-site or at manufacturer's works.

Manufacturers are, of course, responsible for the safe execution of tests carried out at their works, whilst the contractor takes responsibility for the safe execution of all site tests. The purchaser's representative is responsible for the

witnessing of tests but not their execution.

7.1.3 Commissioning Programme

The contractor will be required to draw up a detailed programme for the commissioning of the entire works at an early stage in project construction. The programme will detail the dates for the execution of all tests, including those carried out at manufacturer's works, along with the names of the parties to be represented at each test.

7.2 Existing Hydraulic Circuits and Steam Systems

7.2.1 Cleaning and Flushing

As was mentioned in section 5.3.1, achieving low fouling rates for the heat exchangers of a CHP system will be important to controlling maintenance costs and keeping heat recovery performance close to design. It is, therefore, important that steps are taken to clean up the relevant existing water circuits before the heat exchangers are installed.

These circuits will need to be drained down and temporary hydraulic modifications made to pipework to allow full flushing to be undertaken. The assistance of a proprietary cleaning agent may be required. On satisfactory completion of flushing, the pipework is reconnected and strainers installed. The circuits are then put back in service for a number of weeks, during which time the strainers are regularly cleaned and gradually changed from being fitted with a coarse gauze to a fine one. Once this process is complete, the circuits will be ready to accept the new heat exchangers.

It is important to recognise the requirement for cleaning and flushing of existing hydraulic circuits before work starts on site, to ensure the necessary activities can be incorporated into the programme for the installation works at an early stage. Note should also be taken of the need to take circuits temporarily out of service for flushing.

7.2.2 Maintaining Water Quality

The key to maintaining water quality and cleanliness is to ensure water tightness, thereby minimising the requirement for make-up water.

Existing circuits should, thus, be checked for leaks from valve spindles, flanges, pump shafts etc. and seals re-packed as necessary. Water meters should then be installed on all make-up feeds and water consumption monitored to check that water loss is being controlled to a reasonable level.

In the case of steam systems, the entire water treatment regime should be reviewed to ensure that it meets the requirements of the new heat recovery/waste heat boiler plant to be installed.

7.2.3 Flow Rate Measurement and Balancing

Where orifice plates or pressure tapped valves have not been fitted to the existing circuits to be served by the CHP system, they will need to be installed as part of the project. A means of accurately measuring flow in these circuits is required to enable pump speed to be adjusted, when necessary, to maintain design flow once the additional resistance of the new CHP heat exchangers has been added to each circuit.

7.3 Mechanical Installation

7.3.1 Works Testing

It is faster and more cost effective for functional testing to be undertaken at the manufacturer's works wherever possible. For plant that is assembled on site, such as large waste heat boilers, it will only be possible to test individual components at the works. Full operating tests should, however, be undertaken on all packaged plants before shipping.

In the case of a CHP project, works testing should include:

- Full load tests on the generating plant.
- Full load tests on the packaged fuel gas supply plant.
- Pressure testing of all heat exchangers that are to be shipped as packaged units.
- Logic testing of all control and switching panels.

The full load tests on generating plant and associated control panels should be witnessed by a representative of the purchaser. A written record of the results from all works tests should be kept by the manufacturer for subsequent inspection by a representative of the purchaser.

7.3.2 Functional Testing of Mechanical Infrastructure

Before the main items of a cogeneration plant can be commissioned, the component parts of the mechanical and electrical systems that serve each plant must be tested for functionality to eliminate as many potential problems as possible in advance of commissioning. The electrical testing activities are discussed in section 7.4.2. The functional testing required for the mechanical infrastructure which is to serve the CHP system will include:

- Pressure tests, commissioning and balancing for all water circuits.
- Leak tests for engine intake and exhaust systems.
- Leak tests and commissioning of all mechanical ventilation systems.
- Pressure tests for heat exchangers, waste heat boilers and heat rejection cooling towers.
- Commissioning of all motorised valves and actuators.
- Pressure and leak tests for fuel supply systems.
- Commissioning and testing of fuel conditioning equipment, where installed.
- Commissioning, setting and testing of all mechanical safety devices such as pressure relief valves and slam shut valves.
- The calibration of all sensors and the commissioning of all automatic controls.

Each of the above tests will need to be witnessed by a representative of the purchaser and formal witness test documentation produced.

7.3.3 Technical Inspection

Prior to commencing the commissioning of the major items of plant, a technical inspection of the entire installation should be undertaken to ensure that the works have been fully completed and that all plant can be operated safely. The inspection will include the following:

- Formal confirmation that functional testing has been completed successfully for all mechanical systems by inspection of test documentation.
- Formal confirmation that functional testing has been completed successfully for all electrical systems by inspection of test documen-

tation (see section 7.4).

- A final inspection of the installation works to ensure that materials used, jointing and fixing techniques, bushing arrangements etc. are all in compliance with the specification.
- A final check to ensure that exhaust system condensate pipes are free from debris and water traps are filled.

7.3.4 Gas Compression Plant

The gas compression plant, where installed, can now be commissioned and functionally tested. The engine of the CHP plant will, of course, not be operating at this stage in the commissioning process so the necessary tests have to be completed with minimal compressor running time. The testing of the compression plant will include:

- Correct start up of the plant on the presence of an enable signal, including soft start where fitted.
- Correct shut down of the plant on the removal of the enable signal.
- Correct unloading and shut down of the plant when maximum pressure is reached in the receiver vessel.

7.3.5 Generating Set

Final Checks The next step is to commission and test the generating set. Before this is done, the following final checks should be made:

- Formal confirmation that works testing of the generating set has been successfully completed by inspection of the test documentation.
- Formal confirmation that all generating set control and safety equipment has been correctly installed and functionally tested (as far as is possible without the set running) by inspection of the test documentation.
- A final check to ensure that cooling and lubricating systems are fully charged with water and oil respectively and level control systems are fully operational.
- A final check to ensure that all items of heat recovery equipment are fully charged with water and primary and secondary water circulation is operational.
- A final check to ensure that air or water cool-

ing of the waste heat boost burner is operational.

No Load Tests Once these final checks have been successfully completed, the engine can be fired up and relatively short test runs undertaken at no load to test the following:

- Correct start up of the plant on the presence of an enable signal.
- Correct operation of the fuel supply system and the gas compressor, where fitted.
- Correct operation of the engine cooling and lubricating systems.
- Correct operation of engine speed control.
- Correct shut down of the plant on the removal of the enable signal.
- Correct shut down of the plant in response to all engine safety systems, such as: low oil pressure, high cooling water temperature, low fuel supply pressure and over-speed.
- Correct shut down of the plant in response to a generator circuit breaker trip.
- Correct shut down of the plant in response to the operation of the fire safety system.

Load Tests Once the no load tests on the engine have been completed satisfactorily, then the generating set as a whole can be tested under load. The tests will include:

- Correct operation of alternator AVR control, including measurement of output characteristic and response to a voltage step change.
- Correct operation of automatic synchronising equipment.
- Correct operation of real and reactive load sharing control, where more than one generator is installed.

7.3.6 Waste Heat Boiler

With the engine commissioned and operational, the waste heat boiler and boost burner can be commissioned and tested. Testing will include:

- Correct start up of the boost burner on an enable signal.
- Correct shut down of the boost burner on removal of the enable signal.
- Commission and testing of the operation of automatic controls on the motorised by-pass dampers and the boost burner.

- Commissioning and testing of the operation of automatic blow down controls.
- Setting and testing of safety controls.

7.4 Electrical Installation

7.4.1 Equipment Type Testing and Compliance with National Standards

A schedule will be required detailing all electrical switchgear and protection equipment to be used in the installation. The schedule shall state the manufacturer's component number for each item, the function that the equipment is to serve and the relevant national standard in each case. Finally, copies of certification documents will be required from equipment manufacturers, to demonstrate that all equipment has been type tested to the relevant national standards and meets those standards.

7.4.2 Functional Testing of Electrical Infrastructure

The need for the functional testing of the mechanical and electrical systems which are to serve the cogeneration plant has been discussed is section 7.3.2 and the necessary mechanical testing activities are outlined there. The functional testing of the electrical infrastructure will include:

- Normal electrical testing of all distribution systems, such as continuity, polarity, insulation and earth loop impedance tests.
- Testing of the correct operation and interlocking of all switchgear.
- Earthing system and electrode tests.
- Ratio, vector group and magnetising current tests on power transformers, plus the calibration of winding and oil temperature sensors.
- 10/1 minute polarisation indices on HV rotating plant.

7.4.3 Electrical Protection Tests

A detailed listing and explanation of the commissioning and testing required to set up protective equipment and then prove the correct operation of all protection, is beyond the scope of this book. The following list of activities is provided simply to give the reader an insight into the procedures that will be involved:

- Continuity and resistance tests on protective circuit wiring.
- Magnetisation current/voltage, winding resistance, ratio and polarity tests on current/voltage transformers.
- Vector group and phasing tests on voltage transformer circuits.
- The setting of operating parameters for protective devices in accordance with design requirements.
- Protective circuit operating tests by primary injection, including simulated fault tests, to check the sensitivity and stability of protection systems.
- Protective equipment timing tests by secondary injection.
- The calibration of alarm devices and functional testing of alarms and breaker tripping.

As a guide to the sensitivity and timing tests undertaken on protective equipment, an example test sheet is given in table 7.4(a).

7.4.4 Utility Company Authority

A number of the protective measures required for embedded generation and hence CHP, are installed to protect the supply network of the utility company which serves the site. The utility company will, therefore, require proof that the necessary commissioning and testing of the relevant switchgear and protection has been undertaken. In fact, it is usual for a utility company representative to witness protective equipment testing.

Only when the utility company are fully satisfied that adequate protection of their network is assured will they give authority for the CHP installation to be connected to their supply.

7.5 Overall Plant Performance Tests and Optimisation

Once all the commissioning and testing activities on the individual components of the cogeneration system are complete, then the performance of the overall plant can be tested and adjustments made to optimise that performance. The execution and recording of the various tests to be undertaken are discussed in the sections that follow.

Table 7.4(a)   *Protective Equipment Sensitivity and Timing Tests – Test Sheet*

Item Tested[1]		Results Recorded		
		Operating		Reset
		Value – V/Hz	Time – s	Time – s
Generator Protection				
Reverse power		n/a		n/a
(Unit circulating current)				n/a
Supply Protection – Generator				
Over current				n/a
Phase unbalance				n/a
Over voltage:	R – N or R – Y			
	Y – N or Y – B			
	B – N or B – R			
Under voltage:	R – N or R – Y			
	Y – N or Y – B			
	B – N or B – R			
Over frequency				
Under frequency				
Supply Protection – Intake				
Over current				n/a
Earth fault				n/a
Phase unbalance				n/a
Standby earth fault				n/a
External Supply Protection – Customer				
Over voltage:	R – N or R – Y			
	Y – N or Y – B			
	B – N or B – R			
Under voltage:	R – N or R – Y			
	Y – N or Y – B			
	B – N or B – R			
Over frequency				
Under frequency				
Neutral Earthing[2]				
Neutral earthing circuit breaker close		n/a		n/a

Notes
[1] See figures 3.8(a) and 3.8(b) for the monitoring and control arrangements associated with each item of protection.
[2] The correct operation of neutral earthing circuit breakers in the event of an external supply failure also needs to be tested.

7.5.1 Instrumentation and Test Equipment

Monitored Parameters To generate the data required to evaluate performance, instrumentation will need to be fitted to the plant to measure various critical parameters. The instrumentation typically required for IC engine and gas turbine based CHP plants is detailed in figures 7.5(a) and 7.5(b) respectively.

As the performance of the CHP system has to be evaluated on a regular basis over the lifetime of the installation, it is usual for much of the necessary instrumentation to be permanently installed, rather than just fitted for the initial performance tests. Some items of temporary, portable test equipment will, nevertheless, be needed.

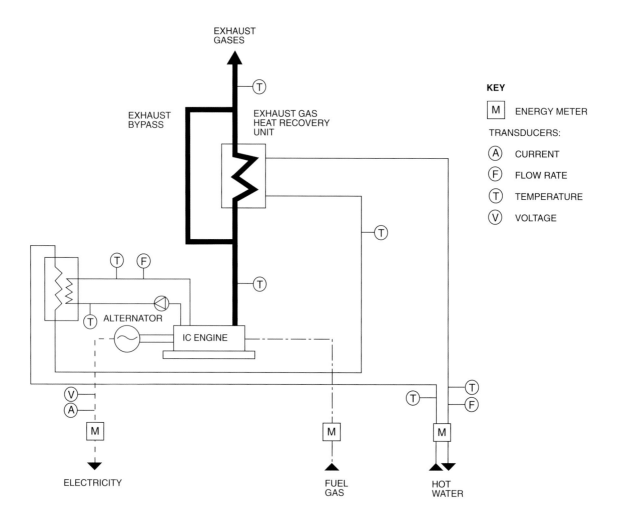

Figure 7.5(a) *CHP System Performance Monitoring Instrumentation – IC Engine based Plants*

Temporary equipment The temporary test equipment required will, typically, include the following items:

- A portable electronic exhaust gas analyser.
- A digital thermometer plus a range of thermocouples suitable for measuring temperature up to 600°C.
- A portable electronic electricity meter plus split CT's.
- A U-tube manometer filled with water.
- A portable sound pressure level meter capable of measuring sound pressure levels in octave bands as well as dB(A).

Calibration certificates for the equipment (where relevant) should be checked to ensure that each item is still within the prescribed calibration period.

Data Logging Until relatively recently, the cost of data logging equipment may have precluded its inclusion as a permanent feature of a CHP installation, with the exception of the logging of perhaps one or two key parameters. This is no longer the case and permanent data logging facilities for all parameters monitored for the purposes of determining plant performance are now included as a standard feature of CHP installations.

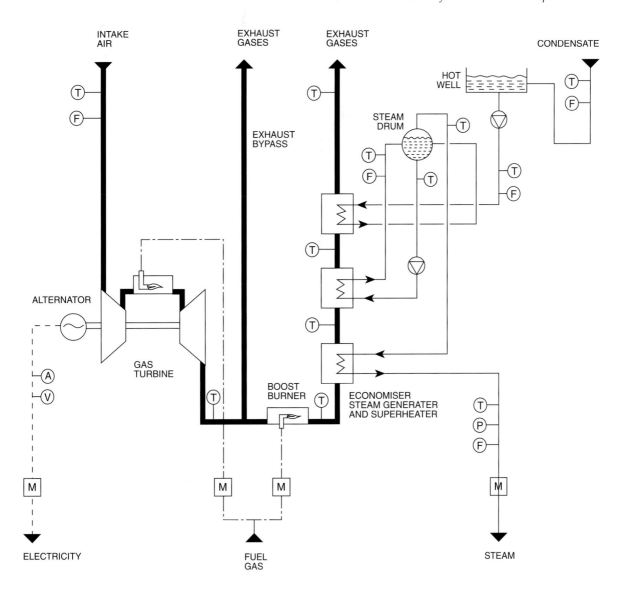

INTAKE
AIR

EXHAUST
GASES

EXHAUST
GASES

CONDENSATE

HOT
WELL

EXHAUST
BYPASS

STEAM
DRUM

ALTERNATOR

GAS
TURBINE

BOOST
BURNER

ECONOMISER
STEAM GENERATER
AND SUPERHEATER

ELECTRICITY

FUEL
GAS

STEAM

KEY

| M | ENERGY METER |

TRANSDUCERS:

(A) CURRENT

(F) FLOW RATE

(P) PRESSURE

(T) TEMPERATURE

(V) VOLTAGE

Figure 7.5(b) *CHP System Performance Monitoring Instrumentation – Gas Tubine based Plants*

Table 7.5(a) *CHP Plant Performance Tests – Electrical Power Output*

Item to be determined	Measurements
Electrical output power	Electrical meter kWh reading taken at the beginning and end of the test
Ambient air temperature	Digital thermometer reading taken at engine air intake
Ambient air pressure	Digital pressure meter reading taken at engine air intake
Engine intake system pressure loss	Differential pressure reading taken using a U-tube manometer
Exhaust system pressure loss	Differential pressure reading taken using a U-tube manometer

Data Presentation The days of endless columns of numbers continuously being dumped to reams of paper are, thankfully, now gone. Logged data is stored on discs for later viewing by plant operators, in tabular or graphical form. Printouts are only produced on request.

The data logging and data presentation facilities are usually provided as an integral part of the overall CHP plant monitoring and control system.

7.5.2 Engine Power Output and Efficiency

The maximum output power from the engine will be measured to check for compliance with the specification. The measurements which will need to be taken are listed in table 7.5(a). Maximum engine output does, however, vary with outside air conditions and intake and exhaust system pressure drop, as discussed in chapter 3. The test results are, therefore, corrected back to ISO standard conditions for engine output which are as follows:

Ambient air temperature: 15°C
Ambient air pressure 1013 mbar
Engine intake system pressure drop: 0 kPa
Engine exhaust system pressure drop: 0 kPa

To illustrate the procedure, corrections to a set of typical test data for a gas turbine are shown in table 7.5(b). Reference should be made to the generating set manufacturer to obtain the necessary correction figures.

Next, the percentage of input fuel converted to electricity and recovered as heat will be determined at 100%, 75% and 50% generating set loading to check for compliance with the spec-

ification. The measurements that will need to be taken and the other input data required are listed in table 7.5(c).

The power generating efficiency of both IC engines and gas turbines varies with ambient conditions. The variation is small, however, with gas turbine efficiency, for example, falling by typically just 0.5% (i.e. 30.0% to 29.5%) for a 10°C rise in ambient air temperature. It is, thus, unnecessary to correct for ambient air conditions unless they are extreme at the time of the tests.

As the performance of steam turbines is independent of ambient air conditions, the correction of test figures is unnecessary for this type of engine.

It should be noted that for power generating efficiency the accepted convention is for manufacturers to give efficiency figures in relation to the lower heat value (lower calorific value) of the input fuel. Care should, therefore, be taken both when specifying and when testing the performance of CHP plant to ensure that the basis for all efficiency figures is clear.

7.5.3 Heat Recovery

The heat recovery performance of the cogeneration plant will also be tested for compliance with the specification. In a similar fashion to power output, test results for engine exhaust gas temperature and mass flow rate, jacket heat rejection and lubricating oil heat rejection can all be corrected back to ISO standard conditions.

Demonstration of expected heat output from an engine does not, however, provide proof that waste heat boilers and heat exchangers have been adequately sized to deliver design heat recovery rates. In practice, this can best be demon-

Table 7.5(b)   *Corrections to a Set of Typical Test Data for a Gas Turbine*

Ref	Item	Manufacturer's Design Performance Data	Site Test Performance Data
Measured Data			
A	Electrical power output – kW	3630	3586
B	Ambient air pressure – mbar	1013	1008
C	Ambient air temperature – °C	15.0	14.2
D	Intake system pressure loss – kPa	1.0	0.7
	– mbar	10	7
E	External system pressure loss – kPa	2.0	0.9
	– mbar	20	9
Correction for Alternator and Gearbox losses			
F	Shaft power correction[1] – kW	172	171
	(A/0.97 + 60 – A)		
Correction for Ambient Air Pressure and Intake Losses			
G	Absolute pressure at gas turbine compressor inlet – mbar	1003	1001
	(B–D)		
H	Compressor inlet pressure ratio (G/ISO ambient air pressure of 1013 mbar)	0.9901	0.9882
I	Shaft power correction – kW	38	45
	([A+F] / H – [A+F])		
Correction for Exhaust Losses			
J	Absolute pressure at power turbine outlet–mbar	1033	1017
	(B+E)		
K	Compressor inlet to power turbine outlet pressure ratio (G/J)	0.9710	0.9843
L	Shaft power correction[2] – kW	77	42
	(2670 × [1 – K])		
Correction for Ambient Air Temperature			
M	Shaft power correction[3]	0	–20
Performance under ISO Conditions			
N	Shaft power – kW	3917	3824
	(A+F+I+L+M)		
O	Percentage of design performance – %	100	98

Notes

[1] From manufacturer's data, alternator efficiency taken as 97% and gearbox losses taken as 60 kW.

[2] From manufacturer's data, compressor inlet to power turbine outlet pressure correction ~ 2670 x (1 – inlet to outlet pressure ratio) for a nominal power turbine inlet temperature of 1035°C.

[3] From manufacturer's performance curve for power versus ambient air temperature.

strated by extrapolation from performance during the tests using the appropriate heat transfer calculations for the components concerned. The alternative is to test the CHP installation at a time of design heat loading and, when ambient conditions permit, design power output from the engine.

7.5.4 Gas Compression Plant

A common problem associated with the operation of gas compression plant is control instability, known as 'hunting'. Such control instability is often simply due to inadequate control loop tuning. In CHP installations, however, experience in the UK has shown that hunting can arise as a result of an interaction

Table 7.5(c) *CHP Plant Performance Tests – Efficiency of Power and Heat Generation*

Item to be determined	Measurements and data required
Fuel input rate	Fuel meter m³/litres reading taken at the begining and end of the test plus fuel calorific value (and pressure for gas)
Electrical output power	Electrical meter kWh reading taken at the beginning and end of the test
Heat recovery to water circuits	Flow and return water temperatures plus commissioned flow rates
Heat recovery to steam systems	Steam meter kg reading taken at the beginning and end of the test plus steam temperature and pressure at meter position

between the modulating control of the gas turbine fuel valve and the gas compression plant loading control. This phenomenon has been discussed in section 5.3.3.

To make sure that hunting does not occur under all likely conditions, temporary testing equipment should be installed to log instantaneous current drawn by the compressor plant motor. Logging should be undertaken over a period of hours with the gas turbine being stopped, started and run at various loads. The logging interval will need to be set at just a few seconds to ensure that any control instability is recorded.

7.5.5 Operating Control Strategy

The automatic controls of a CHP installation need to be checked to ensure that the chosen control strategy is being delivered under all conditions. The best method to check this is to set up data logging for total site electrical and heat demand (derived from fuel consumption if necessary) and total CHP plant electrical and heat output to record the operation of the plant in response to site loads over a number of weeks. The recorded data is then printed out in graphical form and scrutinised to check for correct control strategy operation at various times of the day and for various combinations of site electricity and heat demand.

7.5.6 Exhaust Gas Emissions

Exhaust gas emissions from the CHP plant will need to be checked for compliance with the specification using a portable electronic flue gas analyser. Where the plant is to operate at part load for a significant number of hours per

annum, tests for compliance at part load will also be required.

7.5.7 Noise

The noise in the engine room, at external intake and exhaust positions and at other critical locations in the area surrounding the engine room will need to be taken using a portable sound pressure level meter. The readings taken will then be checked against the specification for compliance.

7.5.8 Performance Optimisation

Non-compliance with the specification in any of the areas detailed in the preceding sections will require the contractor to take action to improve performance where necessary. On a complex plant such as a cogeneration installation, the optimisation process may take a number of months. For this reason, provided that performance is close to design at the time of completion of the works, performance optimisation is normally undertaken during the contract defects liability period.

The performance tests should be repeated approximately 3 months before the end of the normal 12 month defects liability period to check on the degradation of performance over time. If the recorded degradation is within the limits set down in the specification, then no further action will need to be taken by the contractor. Should the rate of degradation exceed the limits specified for the installation, then the contractor will be required to undertake investigations to determine the reason why and implement corrective measures as necessary.

8

OPERATION
and
MAINTENANCE

Like any complex mechanical installation, the operating strategy used to control a CHP plant and the standard to which the plant is maintained will have a major impact on the financial returns produced by the investment.

OPERATION AND MAINTENANCE

8.1 Operating Strategy

8.1.1 The Options

Operating strategy options were touched upon in section 4.5 of chapter 4 with reference to the sizing of generating sets. Detailed consideration of these options is given below.

Maximum Output   In the most simple of strategies, the engine of the cogeneration system is operated continuously at maximum rated output power whenever it is available to run. Electricity generated in excess of site demand is exported and sold to the external electricity supply utility. Heat generated in excess of site demand is dumped.

Electrical Led   In this case, the power output from the engine is controlled to achieve electrical generation equal to the site demand for electricity, up to the maximum continuous rating of the engine. The plant is operated whenever it is available for use and heat generated in excess of site demand is dumped.

Heat Led   Similar to electrical led, the heat led strategy controls the output from the engine to achieve heat recovery equal to the site demand for heat, up to the maximum continuous rating of the engine. The plant is operated whenever it is available for use and electricity generated in excess of site demand is exported and sold to the external electricity supply utility.

Time Dependent Maximum Output/Heat Led
A simple time signal is used to switch this operating strategy from 'maximum output' to 'heat led' at times of low electricity export prices. The logic behind this is that the night-time export prices offered by many electricity utilities are such that controlling the cogeneration system to generate maximum electrical power whilst some heat is dumped, is not economic on an instantaneous operating costs versus revenue basis. The trimming back of power output to a point where all heat is being usefully recovered, can shift the balance of economics back in favour of operating the plant to generate electricity for export.

The plant is, once again, operated whenever it is available for use and, during the day, heat generated in excess of site demand is dumped.

Time Dependent Heat Led/Electrical Led
Similar to the last strategy, in this case a switch is made from a heat led strategy during the day to an electrical led strategy at night when electricity export prices are low. The plant is operated whenever it is available for use.

Time Dependent Heat (or Electrical) Led/Off
The purchase electricity prices available at some sites at night may be so low as to make the generation of electricity on site uneconomic under any operating strategy. This can often be the case with smaller CHP installations where fuel and maintenance costs are relatively high in relation to electrical power output. Under these circumstances, the cogeneration system is simply shut down at night with all electricity being purchased from the external supply utility.

Financial Yield Led   At large sites the application of complex electricity tariffs is common. A feature of such tariffs is the variation of purchase and sale unit charges with the time of day, day of the week and month of the year. Under this type of tariff the economics of CHP plant operation can, of course, vary from hour to hour throughout the year.

In these circumstances, a more sophisticated approach to operating mode determination is required, if the return on investment in the scheme is to be maximised.

Under a financial yield led strategy the cogeneration system is switched between maximum power, electrical led and heat led strategies to achieve the maximum financial yield from the operation of the plant. In addition, where electricity tariffs have punitive peak charges at certain times of the day for certain months of the year, non-combined heat and power emergency generating plant can also be brought on-line under the control of the strategy, to minimise the purchase of electricity at these times.

To choose the optimum operating mode for each hour, the control system has to be programmed with data on the hour by hour purchase prices for the fuels used and sale prices for the heat and electricity generated by the cogeneration system along with the hours run (or kWh generated) component of maintenance charges.

The rapid and frequent cycling of a CHP plant

must, however, be avoided if component life is not to be adversely affected. The control system cannot, therefore, make decisions on an instantaneous basis but must decide on operating mode in advance and apply that decision for a reasonable period of time before switching to an alternative mode.

To take operating mode decisions in advance, the financial yield led strategy has to be able to predict future electricity and heat demands, in addition to knowing about energy and maintenance costs. This type of strategy, therefore, also needs to incorporate self-learning algorithms that can predict demands several hours in advance, based on logged historical data.

Auxiliary Heat Generation   In all the above cases, exhaust gas boost burners and then conventional boilers are operated to generate additional heat when heat recovered from the engine does not meet the demand for heat at the site.

8.1.2 Choice of Control Strategy

General   Which option is best for a particular site will depend very much on the structure and regulation of the electricity supply industry in the country concerned. In countries where the industry is 'vertically integrated', (i.e. single companies generate, transmit and supply electricity) and regulation of the building of new power stations is strict, then export tariffs for CHP installations will generally be favourable. Under these circumstances, maximum output or heat led strategies are likely to be appropriate.

In countries where different companies generate, transmit and supply electricity and regulation is light, leading to the building of significant excess generating capacity, then the export prices that can be secured by small generators will be low, particularly at night. In such situations, electrical led and time dependent strategies will need to be employed.

Bin Calculations
The individual economics that apply at the site in question will, of course, determine which of the simple control strategies will be most cost effective. The procedure for calculating energy cost savings/revenue has been set out in detail in section 4.7. The predictions produced by these calculations coupled with the run time or

electricity generated based maintenance costs for the plant, will enable the optimum operating mode to be determined for the cogeneration system for each time period examined. Armed with this information, a suitable control strategy for the installation can be chosen.

A large cogeneration system installed at a site where complex electricity purchase and sale tariffs apply will almost certainly warrant the selection of a sophisticated financial yield led operating strategy.

Export of Power and Heat   In the case of major combined heat and power schemes whose primary function is to export electricity and heat for sale on a commercial basis, the adoption of a sophisticated financial yield led plant control strategy will be fundamental to achieving the maximum return on the investment.

8.2   Operation

8.2.1 Levels of Automation and the Need for Permanent Site Staff

Automation   Modern CHP plants are fully automatic, with start up, synchronisation, output control and shut down all being undertaken without the need for operator intervention. In addition, the regular monitoring of temperatures, pressures, fluid levels, etc., can all be undertaken automatically and logged for later inspection. Finally, the detection of unusual values for monitored parameters can be programmed into the control system to provide a level of automated condition monitoring for the installation.

Response to Faults   It is in the area of faults or potential faults that human operators still have an important role to play. A manual restart will be required following most plant shut downs that occur as a result of system faults. For a small CHP system, the loss of savings/revenue associated with the loss of service may be too small to justify the provision of technical staff to respond and clear the fault, 24 hours a day. For large installations it will be important that faults are cleared quickly and the system put back into service with the minimum of delay. For such installations, continuous technical staff cover is required. Staff need not, however, remain in permanent attendance of the CHP plant but can execute other duties, provided that:

a) They can be contacted and can respond quickly in the event of a fault.
b) Faults are relayed to a VDU or printer which is permanently manned, by a member of the site security staff for example.

At sites that already have permanent technical staff cover, the additional demands on existing personnel will be minimal.

8.2.2 In-house Maintenance Tasks

In addition to the supervisory duties discussed above, site staff will also be required to undertake routine maintenance tasks on the CHP plant. The extent of these tasks will vary depending on the type of maintenance contract that is purchased for the plant. Typical duties are listed below:

IC Engines
- Replacement of lubricating oil filters and air filters.
- Topping up of lubricating oil and coolant water supply systems.
- Lubricating oil and coolant water changes.
- Dispatch of lubricating oil samples to maintenance contractor for analysis.
- Battery level and acidity checks plus top-ups as necessary.
- Visual inspections of pipes and hoses for leaks.
- Visual inspections of the condition of drive belts plus checks of tensions and adjustments as necessary.
- Attendance on the maintenance contractor's personnel during the execution of major maintenance work.

Gas Turbines
- Weekly on-line cleaning of the gas turbine compressor blades.
- Quarterly off-line soak cleaning of the gas turbine compressor blades at cranking speeds.
- Replacement (or cleaning) of intake air filters.
- Topping up of lubricating oil plus replacement of oil filters and oil changes.
- Attendance on the maintenance contractor's personnel during the execution of major

maintenance work.
- Visual inspections of pipes for leaks.

Steam Turbines
- Topping up of lubricating oil plus replacement of oil filters and oil changes.
- Checks on steam quality at the intake to the steam turbine.
- Attendance on the maintenance contractor's personnel during the execution of major maintenance work.
- Visual inspections of pipes for leaks and steam traps for correct operation.

Boilers and Heat Recovery Equipment
- Normal boiler attendance duties.
- Heat exchanger strainer checks plus cleaning as necessary.

Intakes and Exhausts
- Exhaust system water trap checks plus refilling as necessary.
- Visual inspections of exhaust system for leaks.

Gas Compression Plant
- Topping up of lubricating oil plus replacement of oil filters and oil changes.
- Visual inspections of pipes and hoses for leaks.
- Fuel gas filter checks plus replacement as necessary.

8.2.3 Emergency Procedures

Instruction Manuals and Training   A set of instructions clearly explaining the procedures to be followed in the event of plant shut downs will be produced for the cogeneration system as part of the installation contract. It is important that all staff who are likely to be used to supervise the plant are given comprehensive training on exactly how to react in the event of an outage, in accordance with the emergency procedures.

Manual Restart Following Plant Shut Down
Large engines cost a great deal of money. The total failure of a component can, thus, led to extremely expensive repair bills. For this reason, modern engines are fitted with sophisticated condition monitoring equipment which is designed to detect the potential failure of a com-

1

GAS TURBINE COMPRESSOR FOULING

The Problem In considering the operation and maintenance of gas turbine plant there is one issue, in particular, that demands attention - compressor blade fouling. The effects of gas turbine compressor fouling may include:

- A reduction in maximum power output from the plant of 10% or more.
- A reduction in brake thermal efficiency for the plant of up to 2% (i.e. from 35% to 33%).
- Poor compressor surge behaviour.
- Compressor blade erosion and corrosion which can result in early blade failure.

The rate at which fouling occurs in a compressor will be dependent upon the prevailing weather conditions and outdoor air quality amongst other factors. In spite of this fact, many operators of gas turbines use an essentially time based approach to compressor cleaning. Whilst such regimes can be modified to suit a particular installation over a number of years and may incorporate seasonal and other operating factors, a far more effective approach is to base cleaning on plant condition monitoring.

Fouling Detection The three most common techniques used to detect the on-set of gas turbine fouling are:

1. The regular calculation of compressor efficiency from measured operating parameters.
2. The continuous monitoring of intake air mass flow rate using an intake depression manometer.
3. Regular visual inspections of compressor blades.

Technique 1 provides the most direct indication of the effect of compressor fouling. The accuracy with which the required operating parameters (power output, exhaust gas temperature and compressor discharge temperature and pressure) have to be measured, however, makes this technique difficult to apply reliably in practice.

Regular visual inspections can provide a sensitive and reliable means of detecting the on-set of compressor fouling. The disadvantages are that an expert technician is required to undertake the inspections and, in the case of some designs of gas turbine, the plant has to be shut down for inspection hatches in the compressor casing to be removed to provide the necessary access.

The use of pressure tappings at two points along the bell mouth of the intake to a compressor can be used to give an indication of air velocity and hence mass flow rate into a gas turbine. As the pressure difference between the tappings is approximately proportional to the square of velocity, the technique is highly sensitive to changes in air mass flow rate and hence compressor fouling. In addition, as the technique requires the measurement of one simple parameter, it is a relatively straightforward matter to provide automatic indication of unacceptable compressor fouling.

Air Filtration The primary method of controlling fouling has, of course, to be intake air filtration. This has been discussed in section 5.2.3 of chapter 5. Whilst the choice of filtration technique is, perhaps, the primary consideration a number of other important issues must not be overlooked:

- The air tightness of the complete intake system. Leaks through poorly finished ductwork joints, test holes and corrosion perforations will negate the effectiveness of the most comprehensive filtration system.
- Air leakage around badly fitting or through open filter by-pass doors. By-pass doors are installed to provide an additional path for air flow in the event of an unacceptable rise in pressure drop across the intake filter system. Degradation of seals over time may cause the door to leak whilst the failure of the automatic door opening mechanism can lead to the door remaining permanently ajar.
- Flange and access door distortion under op-

erating depressions. The forces applied to ductwork and access doors as a result of the sub-atmospheric pressure developed within the intake system can lead to the distortion of components and the creation of leaks that would not occur under the positive pressures that may be used for testing.

- The uniformity of air velocity across the filter surface. For maximum filtration efficiency, intake ductwork leading up to filters should be designed to provide a uniform flow of air across the entire surface of the filter element.

Cleaning   There are two approaches to the cleaning of compressor blades. In the first approach common abrasive materials, such as ground nutshells or rice, are introduced into the air intake to the gas turbine when the machine is operating on-load at full speed. Whilst particularly effective for the removal of hard dry deposits, this approach can cause serious problems when used in modern designs of machine. These problems include the erosion of protective blade coatings and the clogging of the fine air-ways now used to cool the downstream turbine blades.

The second approach is to use purified water with or without the addition of detergents. The water may be introduced via atomising nozzles whilst the turbine is on-load and at full speed. In this case, the cleaning action is essentially one of physical water impact. When detergents are used, the gas turbine is usually operated at cranking speed enabling a chemical cleaning action to take place. This type of 'soak wash' is followed by a number of water rinse cycles where the gas turbine is brought up to approximately half starting speed then allowed to coast to a stop.

Advice on the optimum cleaning regime for a gas turbine must, of course, be sought from the manufacturer of the machine. A typical programme will include regular weekly on-line water washes with soak washes at cranking speeds every month or so.

Optimisation of the Cleaning Regime   Soak washes require a gas turbine to be off-load and hence not generating heat and power. The maintenance of compressor performance has, therefore, to be a trade-off between efficiency and plant down-time. Where condition monitoring is used to detect the on-set of compressor fouling, the frequency with which soak washes are undertaken can be varied to suit prevailing weather and air quality conditions. In this way, the optimum balance can be struck between the need for cleaning and the requirement to maximise plant operating hours.

ponent, such as a bearing, in advance of actual failure and shut the plant down. It will be realised, therefore, that only manual restart of the engine can be allowed following a plant shut down of this nature.

Most manufacturers will, nevertheless, have stories of incidents when over-confident plant operators have restarted engines following a vibration induced shut down, without bothering to investigate the fault. The results have included total bearing failure, damage to drive shafts, and the destruction of gear boxes.

It is vital, therefore, that operators always refer to the instruction manual in the event of an emergency shut down, to determine the appropriate course of action. In many cases, a telephone conversation with the manufacturer will be necessary to decide whether or not the engine can be restarted without further detailed investigation work.

The lesson here is that, whilst site staff must be confident in the supervision of a cogeneration system, a clear distinction must be made between the operation of normal building services plant and a combined heat and power installation.

8.2.4 Management

For major schemes, particularly where significant export of electricity and/or heat is to be undertaken, a member of staff of the organisation responsible for operating the CHP installa-

Table 8.3(a)   *Indicative Maintenance Requirements – IC Engine Operating on Natural Gas*

Elapsed Operating Time – hours	Engine Components	Maintenance Work
1,000–2,000 and every 1,000–2,000 thereafter	Lubricating oil systems	Send oil samples for analysis Check oil pressures and levels Replace main engine oil depending upon findings from oil analysis Replace turbo-charger oil depending upon findings from oil analysis Visually inspect sealing rings
	Fuel supply and carburation systems	Measure exhaust oxygen Check and clean air filters Check and clean fuel gas filters Check tightness of throttle valve Visually inspect control rods
	Engine block, cyclinders and cylinder heads	Measure compression Check and reset valve clearances
	Cooling systems	Check levels Visually inspect hoses Check operation of pressure relief valve
	Ignition	Clean and adjust spark plugs Check distributer Check and adjust timing
	Drive and timing belts	Visually inspect belts for condition Check and adjust belt tensions
4,000–6,000 and every 4,000–6,000 thereafter	Lubricating oil systems	Replace oil filters
	Fuel supply and carburation systems	Replace air filters Clean fuel gas filters and replace as necessary Overhaul turbo-charger and replace bearings Clean and inspect charge air cooler
	Engine block, cylinders and cylinder heads	Remove cylinder heads, clean out water spaces, examine valves, springs, rocker gear etc Clean camshaft bracket, camshaft bearings, tappet and oil pipe assemblies and check wear Examine large end bearings and main bearings, securing nuts and split pins. Check bearing clearances
	Cooling systems	Reverse flush water spaces in cylinder casings
	Ignition	Replace distributer cover Replace spark plug covers
	Controls	Check correct operation of all control and safety systems

Table 8.3(a) *continued*

Elapsed Operating Time – hours	Engine Components	Maintenance Work
8,000–12,000 and every 8,000–12,000 thereafter	Lubricating oil systems	Dismantle and clean lubricating oil pumps
	Engine block, cylinders and cylinder heads	Withdraw pistons and clean
		Check that rings are free and all clearances are within tolerances
		Inspect crankcase pins and journals for wear and check clearances
	Cooling systems	Inspect all cooling systems for scale formation and descale components as necessary
	Exhaust system	Check manifolds, silencers and pipework for corrosion and replace components as necessary
16,000–24,000	Engine block, cylinders and cylinder heads	Replace exhaust valves and guides
		Replace main and large end bearings
	General	Replace other components as necessary

tion will be charged with responsibility for the overall management of the plant. This has been touched upon in section 5.10.

This responsibility will include the day-to-day management of the operation and maintenance of the plant but will also extend to optimising the performance of the system and maximising the financial returns achieved by the scheme.

8.2.5 Contracting Out the Operation of Plant

From the above, it will be seen that whilst modern plants are fully automated and can be run unattended the operating duties, particularly for large plants, can be onerous. For this reason, if suitable technical staff are not already employed at a site then it may be appropriate to consider contracting out both the maintenance and the operation of the CHP plant and other heat generating plant at the site.

8.3 Maintenance

8.3.1 Requirements

The maintenance activities that are associated with typical engines are outlined in tables 8.3(a), 8.3(b), and 8.3(c) for IC engines, gas turbines and steam turbines respectively. As has been stated throughout the book, the maintenance of even small IC engines is a specialised activity that is unlikely to be undertaken cost effectively in-house. This section does not, therefore, cover the maintenance that is needed on each type of engine in great detail. Listed below, however, are items to look out for in the work carried out by others, to check that adequate maintenance is being undertaken:

IC Engines
- Are performance tests being undertaken regularly to check on power generating efficiency?
- Are exhaust gas tests being undertaken regularly as a check on carburation settings?
- Are cylinder compression tests being undertaken regularly as a check on cylinder lining, piston ring and valve wear?
- Is sufficient attention being paid to the maintenance of fuel conditioning equipment such as filters, driers and centrifuges?
- Is the operation of pressure relief valves checked annually?
- For spark ignition engines, is electronic diagnostic equipment used regularly to optimise the performance of the ignition system, including timing?
- Is there any evidence to suggest that mainte-

Table 8.3(b) *Indicative Maintenance Requirements – Gas Turbine Operating on Natural Gas*

Elapsed Operating Time – hours	Engine Components	Maintenance Work
–	Compressor blading	Wash on-line and soak wash at cranking speeds at intervals determined by condition monitoring
1,000	Combustion chambers plus compressor and turbine blading	Visually inspect using boroscope
2,000 and every 2,000 thereafter	Lubricating oil system	Send oil sample for analysis and check level
	Intake air systems	Check intake air filters
7,000	Comustion chambers plus compressor and turbine blades	Visually inspect using borescope
8,000	Combustion chamber	Dismantle and inspect
	Controls	Check correct operation of all control and safety systems
	Starter, auxiliary gearbox, turbine, main gearbox and alternator	Check alignments
	All ancillary equipment	Visually inspect
	Lubricating oil system	Change oil and oil filters as necessary
	Intake air system	Change intake air filters as necessary
15,000	As 7,000 hours	
16,000	As 8,000 hours	
23,000	As 7,000 hours	
24,000	As 8,000 hours plus	
	Compressor section	Dismantle and inspect all components and check clearances
	Turbine blades	Remove sample blades for off-site condition assessment
	Auxiliary gearbox	Dismantle and inspect all components and check clearances
31,000	As 7,000 hours	
32,000	As 8,000 hours	
39,000	As 7,000 hours	
40,000	As 24,000 hours plus	
	Turbine section	Dismantle and inspect all components and check clearances
	Reduction gearbox	Dismantle and inspect all components and check clearances
	General	Replace ancillary equipment as necessary

Table 8.3(c)   *Indicative Maintenance Requirements – Steam Turbine*

Elapsed Operating Time – hours	Engine Components	Maintenance Work
2,000 and every 2,000 thereafter	Lubricating oil system	Send oil sample for analysis and check oil level
8,000 and every 8,000 thereafter	Turbine section	Inspect and check clearances on rotor journal and thrust bearings Clean air seal blower strainer
	Lubricating oil system	Change oil and oil filters as necessary
	Controls	Dismantle and clean speed governor and emergency stop valve Check correct operation of all control and safety systems
	Cooling systems	Clean water sides of condenser and oil cooler
	Reduction gearbox	Inspect and check clearances on journal and thrust bearings
24,000	Turbine section	Dismantle and inspect all components and check clearances Replace blades, bearings and labyrinth packings as necessary Dismantle and inspect air pump and extraction pumps
	Lubricating oil system	Dismantle and inspect oil pump Chemically clean oil side of oil cooler
	Reduction gearbox	Dimantle and inspect all components
	General	Replace ancillary equipment as necessary

nance work is being altered in response to results from the analysis of lubricating oil samples sent from site?

• Where automated condition monitoring and recording facilities are provided, is there any evidence that these are being examined, perhaps on a monthly basis, and action being taken in response to the findings?

• Is the engine being stripped down for the visual inspection of bearing surfaces, the checking of clearances etc. in exact accordance with the manufacturer's published recommendations?

• Are parts being replaced, regardless of apparent condition, in exact accordance with the manufacturer's recommendations?

• Is sufficient attention being paid to the maintenance of the correct water treatment and level of cleanliness in the engine jacket cooling circuit?

• Is oil consumption being monitored and regularly checked against manufacturer's data?

Gas Turbines

• Are performance tests being undertaken regularly to check on power generating efficiency?

• Are exhaust gas tests being undertaken regularly as a check on combustion control settings?

• Is sufficient attention being paid to the maintenance of fuel conditioning equipment such as filters, driers and centrifuges?

• Is the data recorded by condition monitoring equipment being examined on a monthly basis and action being taken in response to the findings?

- For two fuel gas turbines, is an automatic fuel gas purge of oil injection nozzles undertaken each time the set is switched over from oil to gas operation, to minimise the build up of carbon deposits in the nozzles?
- Have samples of critical components, such as turbine blades, been removed for detailed condition evaluation at the manufacturer's works after 2 to 3 years of engine operation?

Steam Turbines
- Is sufficient attention being paid to combustion efficiency and, in particular, to the maintenance and sensor calibration of oxygen trim control systems on the steam raising boilers?
- Is adequate attention being paid to water quality, blow down control and the correct operation of make-up water conditioning equipment?
- Are performance tests being carried out regularly to check on the power generating efficiency of the turbine?
- Is the data recorded by condition monitoring equipment being examined on a monthly basis and action being taken in response to the findings?
- Is the operation of pressure relief valves being checked annually?

Boilers and Heat Recovery Equipment
- Is the heat recovery performance of heat exchangers and waste heat boilers tested on a regular basis?
- Are heat exchangers regularly dismantled and both heat transfer surfaces thoroughly cleaned, in response to the results of performance tests?
- Are the heat transfer surfaces of waste heat boilers regularly and thoroughly cleaned, in response to the results of performance tests?
- Is adequate attention being paid to water quality, blow down control and the correct operation of make-up water conditioning equipment on steam raising boilers?
- Is the operation of pressure relief valves checked annually?

Intakes and Exhausts
- Where automatic roll filters or automatic fil-ter cleaning equipment is installed, is it being adequately maintained?
- Are exhaust system condensate drains being regularly cleared and U-traps checked for water?
- Is the exterior of the exhaust ductwork inspected for signs of corrosion or physical weakness annually and sections replaced, particularly bends, as necessary?
- Are the exteriors of flexible connections visually inspected for signs of failure annually?

Gas Compression Plant
- Is lubricating oil use monitored and regularly checked against manufacturer's data to ensure that any fall off in oil separator effectiveness is detected promptly?
- Is the operation of the capacity control of the plant studied for a period of time, or better still data logged, on a regular basis to check for unnecessary compressor cycling?
- Is the operation of pressure relief valves and slam shut valves checked annually? In addition, is the gas tightness of safety shut off valves checked annually?

Alternators
- Is the data recorded by condition monitoring equipment being examined on a monthly basis?

Electrical Switchgear and Protection
- Is the testing of protection equipment undertaken regularly in accordance with the requirements of the external supply utility and local regulations/codes of practice?
- Is the correct operation of automatic synchronising equipment tested regularly?

Fire Safety Systems
- Are fire detection and fuel supply shut-off systems tested annually?
- Are interlocks with ancillary plants such as ventilation fans tested annually?

8.3.2 Terms and Conditions for Maintenance Contracts

The standard terms and conditions offered by manufacturers will, of course, vary widely from company to company. It is vital, therefore, that the purchaser reads and understands these

2

CONDITION MONITORING

Background

Condition monitoring is becoming an ever more important feature of the maintenance of mechanical plant. Conventionally, in the maintenance of any item of equipment the operator has two options:

a) To change components well within their expected operating life to minimise the risk of failure.

b) Only to change components in the event of failure.

Whilst option b) may be satisfactory for small, simple items such as pumps and fans, the expense associated with the complete failure of a component makes this option unacceptable for complex plant. For these reasons option a) has, in the past, been adopted for the maintenance of engines.

As more is learned about materials, metal fatigue and corrosion, however, it has become possible to gauge the further life of a component through the use of condition monitoring techniques. This third maintenance option, of replacing parts before failure but only when they are showing signs of deterioration, is now widely applied to the maintenance of engines and, in particular, to the maintenance of gas and steam turbines.

Techniques

Visual Inspection In the main, condition monitoring involves the visual inspection of components by highly experienced maintenance personnel, to check for signs of fatigue or unacceptable wear as indicated by excessive clearances between components. To reduce the need for stripping down, boroscopes are used extensively in the visual condition monitoring of gas and steam turbines.

Oil Analysis The science of lubricating oil analysis is rapidly becoming more and more important as a monitor for condition in IC engines in particular. In the laboratory testing

of oil there are 3 primary areas of interest:

a) The type and quantity of the materials accumulated in the oil as a result of the wear of components.

b) The presence of water and glycol in the oil.

c) The chemical and physical condition of the oil.

The type and quantity of the materials present in a lubricating oil sample can, when studied by an experienced tester, give an early indication of the incipient failure of a bearing or excessive piston ring wear, to give two examples. A relatively high level of water and glycol in the oil, on the other hand, indicates the failure of gaskets or the possibility of a hairline fracture in an engine block. Finally, knowledge of the rate of degradation of the chemical and physical condition of the oil, gained from the study of successive samples over a period of time, will enable a tailored oil change schedule to be drawn up and implemented for the engine concerned.

Temperature and Vibration Detection The latest condition monitoring techniques utilise the continuous sensing of temperatures and vibrations to build up data on the normal operation of machine components. For example, vibration measurement on a shaft may produce a complex trace of oscillations of different harmonics of rotational speed. This trace or 'signature' will be repeated again and again over time, provided that conditions do not change. The impending failure of a bearing, however, will at first cause the slightest of changes to this trace of oscillation. As the bearing gets closer and closer to failure, the trace will diverge further and further from the original recorded signature. The idea is that a regular review of the trace produced for the vibration monitored will enable an expert analyst to detect the impending failure of the bearing, almost as soon as it starts to degrade.

This same technique can be applied to the temperature and vibration monitoring of any component in an item of mechanical plant. For this approach to succeed and be worthwhile, however, three things have to happen:

a) Normal signatures for all components monitored have to be recorded and stored for later use.

b) Traces for each monitored component must be analysed and compared with the original signature at regular intervals.

c) The level of expertise of the analyst must be sufficient for impending failures to be detected before they could have been detected by other means, by oil analysis for example.

Great claims are made by some manufacturers for this type of automated monitoring technique which can, of course, be undertaken remotely with the assistance of an autodial modem and a telephone line. Failure to perform in one of the three areas outlined above, however, often means that automated monitoring does not deliver all that it is capable of. Reliance on this type of monitoring technique to the exclusion of conventional visual inspections and oil analysis is, therefore, not recommended.

terms completely to ensure that they meet the requirements of the installation in question. These standard conditions should not be considered to be cast in stone and the purchaser should insist on modifications where necessary and reasonable.

The issues that, in particular, will need to be addressed by the maintenance agreement are:

- The length of the agreement.
- The basis on which charges will be made i.e. plant operating hours, kWh of electricity generated or a combination of the two.
- The charges and the price indexing arrangements over the duration of the agreement.
- Exactly what is and is not included in terms of maintenance activities.
- Exactly what is and is not included in terms of spare parts and consumables.
- The duties that are expected of the purchaser i.e. taking oil samples, returning logged data and undertaking simple maintenance tasks.
- The tools, consumables and spare parts that are expected to be purchased and carried by the purchaser.
- The details of the availability guarantee provided by the manufacturer i.e. minimum availability percentage, penalty charges to the paid by the manufacturer on failure to meet the minimum percentage in one year, the bonus charge to be paid to the manufacturer when minimum availability is exceeded in one year, exclusions.

- The provision of loan components in the event of a major failure.
- Guaranteed response times to breakdown call-outs.
- Lifting gear and any other facilities that are to be provided by the purchaser for the purposes of engine maintenance.
- The number and grade of the purchaser's personnel that are required by the manufacturer to assist with the maintenance work and the number of man hours that will be required per annum.

8.4 Performance Monitoring

8.4.1 Requirements

Cogeneration plants comprise a number of comlex components whose performance will deteriorate rapidly should the necessary operation and maintenance work not be undertaken correctly at the prescribed intervals. The key to achieving high levels of performance from CHP systems is, therefore, continuous monitoring.

The performance monitoring instrumentation typically required for IC engine and gas turbine based plants is detailed in figures 7.5(a) and 7.5(b) of chapter 7, respectively. It is, of course, important to ensure that the calibration of all monitoring and metering equipment is checked and re-calibration undertaken as necessary, at least once a year.

8.4.2 Reporting

No matter how sophisticated monitoring instrumentation may be, operating benefits are only obtained where formal procedures are in place for the regular reporting and review of performance data. Experience in the UK and elsewhere has conclusively demonstrated that adequate performance monitoring and the implementation of action in response to performance reports will substancially improve the financial performance of a cogenerating scheme.

Appendix 1

WORKED EXAMPLE OF A PRELIMINARY APPRAISAL

To illustrate the approach set out in section 2.6 to the Preliminary Appraisal of the potential for CHP at a site, a worked example is given here.

APPENDIX 1: WORKED EXAMPLE OF A PRELIMINARY APPRAISAL

A1.1 Site Description

General   The site in question is a medium sized hospital which provides a full range of medical facilities. The accommodation is comprised of one major building and a number of smaller satellite buildings distributed across a 500,000 m² estate.

Electrical Supply Arrangements   Electricity is purchased at high voltage from the local supply utility. A high voltage network is used to distribute power to local sub-stations where the electricity is transformed down to low voltage for final distribution.

Heat Supply Arrangements   Natural gas is purchased under an 'interruptible' supply contract for use in conventional fire tube water boilers. The 4 boilers generate medium temperature hot water (MTHW) at 120°C for distribution throughout the site. The heating and hot water service requirements at each building are served from the MTHW mains via local MTHW to low temperature hot water calorifiers. Where required, steam is generated locally.

Heavy fuel oil is used as the standby fuel in the event of gas supply interruption.

A1.2 Available Energy Data

At the site the following data is available for the preliminary evaluation:

- Monthly fuel bills for gas and electricity for one year.
- One week of hourly electrical demand data.
- Gas meter readings taken once every 4 hour shift for a number of typical weekdays and weekend days.
- The annual overall generating efficiency of the boiler plant is estimated to be 70%. As is usual for boiler plant, the figure for efficiency is given with reference to the *higher heat value* of the fuel used.
- The current energy unit prices for electricity throughout the year are: £0.0486/kWh from 07.00 hours to midnight, Monday to Friday; £0.0234/kWh at all other times.
- The current energy unit price for natural gas throughout the year is £0.0085/kWh (at *higher heat value*).

Table A1.2(a)   *Data from Monthly Fuel Bills*

Month	Electricity		Gas	Heat[1]
	Consumption – MWh	Maximum Demand – kW	Consumption – MWh	Consumption – MWh
October	537	1,550	2,572	1,800
November	636	1,480	3,450	2,415
December	685	1,420	3,102	2,171
January	582	1,420	2,231	1,562
February	655	1,380	2,515	1,761
March	588	1,400	2,512	1,758
Winter Totals	*3,683*	*–*	*–*	*11,467*
April	467	1,360	1,858	1,301
May	584	1,340	1,274	892
June	486	1,340	1,471	1,030
July	483	1,320	1,149	804
August	625	1,420	1,724	1,207
September	532	1,480	2,024	1,417
Summer Totals	*3,177*	*–*	*–*	*6,651*

Notes
[1] Figures for heat simply derived from gas consumption multiplied by annual overall boiler plant generating efficiency.

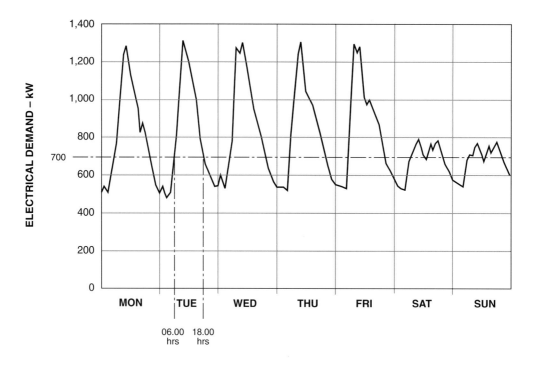

Figure A1.2(a) *Hourly Electrical Demand Data for a Typical Week*

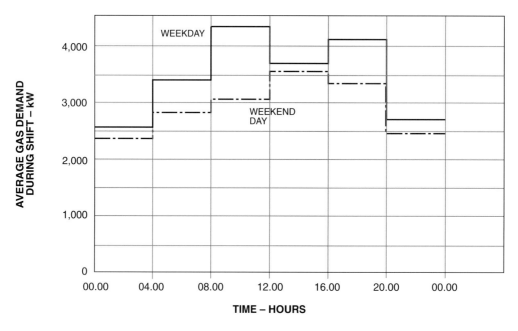

Notes

1. Gas demand given in kW at *higher heat value*.

Figure A1.2(b) *Four Hourly Gas Demand Data for Two Typical Days*

The consumption data is summarised in table A1.2(a) and figures A1.2(a) and A1.2(b).

A1.3 Analysis of Energy Data

Electricity From an inspection of the fuel bills, 'winter' has been taken nominally as running from October to March and 'summer' from April to September. The number of weeks in both summer and winter is, therefore, 26.

From an inspection of the available hourly electrical demand data, 'major use hours' appear to occur nominally from 06.00 to 18.00 hours from Monday to Friday. All other times will, therefore, be taken as being 'out of hours'. From this, the number of hours per week defined as 'major use hours' is 60 and as 'out of hours' is 108.

Again from inspection of the hourly electrical data, the average demands for the 'major use hours' and 'out of hours' periods are approximately 1,000 kW and 600 kW respectively. These figures must, however, be modified to represent the variation in demand between winter and summer. The maximum demand figures taken from the monthly fuel bills should be used as a guide to the likely variation in 'major use hours' average demand. Homing in on representative figures then becomes a process of engineering judgement and iteration using the following steps:

- Initial figures for 'major use hours' and 'out of hours' average demands are chosen.
- These figures are then multiplied by the appropriate number of hours in the week and weeks in the year for each period.
- The total of 'major use hours' and 'out of hours' consumption for winter/summer can then be compared with figures for winter/summer from the fuel bills.
- Adjusted figures for average demand are then chosen.

Table A1.3(a) *Site Demand Patterns – Worked Example*

Bin	Description	Number of Weeks	Hours per Week	Average Demand – kW		Consumption – MWh	
				Electricity	Heat	Electricity	Heat
1	Winter – Major use hours	26	60	1,150	3,100	1,794	4,836
2	Winter – Out of hours		108	650	2,300	1,825	6,458
Total consumption for Bins 1 and 2						3,619	11,294
Total consumption from winter fuel bills						3,683	11,467
3	Summer – Major use hours	26	60	1,000	1,500	1,560	2,340
4	Summer – Out of hours		108	600	1,500	1,685	4,212
Total consumption for Bins 3 and 4						3,245	6,552
Total consumption from summer fuel bills						3,177	6,651

This process is repeated until a reasonable agreement is reached. A photocopy of table 2.2(a) from chapter 2 will be of assistance.

The final estimates of average electrical demand determined for this site are presented in table A1.3(a). The fact that 26 plus 26 weeks, times 7 days per week multiplies up to 364 days not 365 is ignored.

Heat The winter and summer 'major use hours' and 'out of hours' time periods have been determined already under the electrical analysis, as electricity is the dominant consideration in any CHP evaluation.

From inspection of the 4 hourly gas demand data, average demands for the defined 'major use hours' and 'out of hours' time periods are 4,000 kW and 3,000 kW of gas respectively (at higher heat value). Converted to heat demand by multiplying by annual overall generating efficiency, the figures become 2,800 kW and 2,100 kW.

The procedure for modifying these figures to represent the variation of demand between winter and summer is as described under 'Electricity' above. This time, however, there are no monthly maximum demand figures to be of assistance. A knowledge of the number of boilers typically required in winter and summer has to be used instead to 'guesstimate' the seasonal variation in the demand for heat.

The final estimates of average heat demand determined for this site are presented in table A1.3(a).

A1.4 Selection of Engine Technology

With the demand figures needed for table 2.2(a) compiled, the next step is to determine the likely engine technology for the site. To do this, use should be made of the flow chart presented in figure 2.6(a) of chapter 2.

A course plotted along the flow chart for this site is shown in figure A1.4(a). The chart suggests that the correct choice of engine technology for the site is an IC engine.

A1.5 CHP Plant Size

From section 2.6 of chapter 2, it will be remembered that a number of different CHP plant sizes may need to be tried, even at the preliminary evaluation stage, to achieve an economic proposition for cogeneration at a site. The section also advises that a good size to try first is a plant electrical output equal to the overall average electrical demand for the year. For this site, from table A1.2(a), the relevant figure is 783 kW so an 800 kWe CHP plant will be evaluated first.

A1.6 Site Demand Displaced by CHP Plant

The first step in determining the demand displaced by the CHP plant is to find the maximum rate at which heat can be recovered from the engine for the chosen size and technology, when it is operating at full power. This can be read in terms of 'specific heat recovery rate' (kW thermal per kW of electrical output) from figure 2.6(b) of chapter 2. Figure A1.6(a) shows the relevant line for this evaluation drawn onto the graph, giving a figure of 1.40 kWt/kWe for the chosen engine.

The design electrical and heat output capacities of the CHP plant to be tested are now known. With these maximum figures to hand, the actual displacement of electrical site demand and then site heat demand can be determined for each of the four time periods under consideration. The procedure, which makes use of table 2.6(a) of chapter 2, is described in detail in the notes accompanying that table.

The calculated displaced demands for this site are presented in table A1.6(a).

A1.7 CHP Plant Fuel Demand

The 'specific fuel consumption rate' (kW of fuel at lower heat value per kW of electrical output) for the chosen engine at full load can also be read from figure A1.6(a). For this evaluation, the rate is 2.90 kWf/kWe. Correction factors for part load operation will also need to be read from the relevant plot in figure A1.6(a) to give the actual fuel consumption rate, at the relevant electrical power output, for each of the four time periods.

Again, the calculated figures for this site are presented in table A1.6(a).

A1.8 Energy Charges

Electricity Unit Prices Where electricity unit prices vary depending upon the time of day and day of the week, the various charge periods are unlikely to exactly match the 'major use hours/out of hours' calculation periods chosen for the site. For the purposes of this preliminary

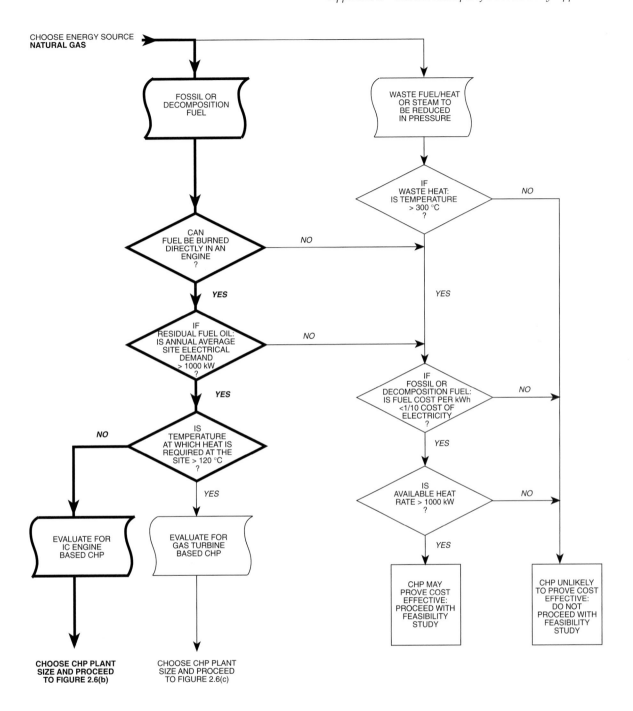

Figure A1.4(a) *Flow Chart for the Preliminary Evaluation of CHP Scheme Potential – Worked Example*

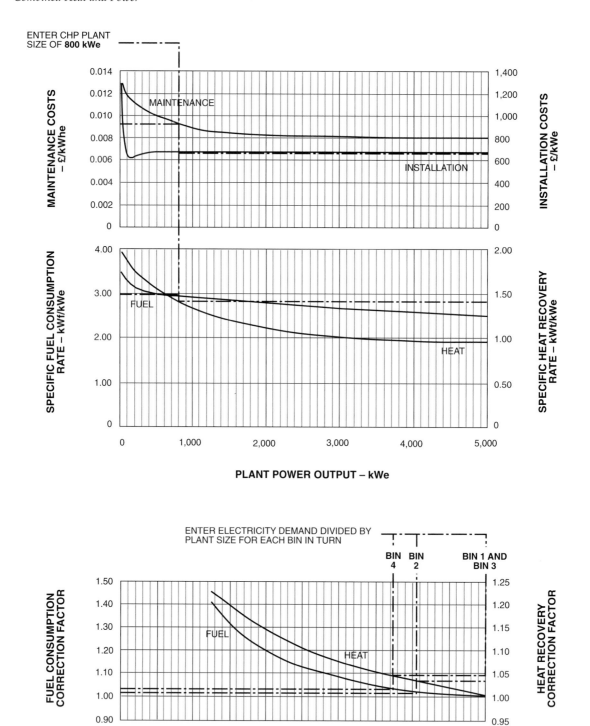

Figure A1.6(a) *Nomograph for the Preliminary Evaluation of CHP Scheme Potential – IC Engines Worked Example*

Table A1.6(a)   *Table for the Preliminary Evaluation of CHP Scheme Potential – Worked Example*

Bin	CHP Plant Running Hours	Demand Displaced by CHP Plant – kW		CHP Plant Fuel Demand – kW	Average Energy Unit Prices – £/kWh			CHP Plant Maintenance Costs – £/kWhe	Revenue Savings for the Scheme – £pa
		Electricity	Heat		Electricity	Heat	CHP Fuel		
	A	B	C	D	E	F	G	H	I
1	1,404	800	1,120	2,320	0.0465	0.0121	0.0094	0.0093	30,192
2	2,527	650	937	1,923	0.0383	0.0121	0.0094	0.0093	30,606
3	1,404	800	1,120	2,320	0.0465	0.0121	0.0094	0.0093	30,192
4	2,527	600	877	1,806	0.0383	0.0121	0.0094	0.0093	27,886
Totals	7862								118,876

CHP Plant Installation Costs – £/kWe	CHP Plant Size – kWe	Total Capital Cost for the Scheme – £
J	K	L
660	800	528,000

evaluation adequate accuracy will be achieved by simply computing the weighted average unit price for electricity, for the 4 calculation periods under consideration.

Electricity Standing and Maximum Demand Charges It should be noted that the unit prices used for electricity must exclude standing and maximum demand charges. At the preliminary evaluation stage, any maximum demand charge savings that might be achieved by the CHP plant are ignored.

Engine Fuel Price and Lower Heat Value As discussed in insert panel 6 of chapter 3, the performance of engine plant is usually given in terms of the lower heat value of the fuel concerned. This convention is, therefore, used in the graphs of figures 2.6(b) and 2.6(c) of chapter 2. The price for engine fuel used in the calculations must, therefore, be in terms of cost per unit of energy at *lower heat value*. For natural gas, the ratio of lower heat value to higher heat value can be taken as 0.9:1. Hence, the cost of gas for use in the engine of the CHP plant in this case is £0.0094/kWh.

A1.9 Maintenance Costs
An approximate figure for the maintenance costs that would be associated with the proposed CHP plant can be read for the chosen engine technology and size from figure A1.6(a). The costs are given in terms of £ per kWh of electricity generated for CHP plants that operate continuously at full load.

Maintenance costs are higher for plants that run for less than 24 hours per day or plants that are operated at part load. Cogeneration installations are, however, generally sized to operate at full output for most of the year and must record a high number of operating hours per annum to achieve a satisfactory return on capital. For this reason, maintenance costs for CHP plants will not, in practice, be significantly affected by operating hours and loading considerations. For the purposes of a preliminary evaluation these considerations can, therefore, be ignored.

For the CHP plant being examined under this evaluation, the figure for maintenance costs is £0.0093/kWhe.

A1.10 Capital Costs
A guide to the likely installation costs (including investigation, design and project management fees) that would be associated with the CHP plant can be read from the relevant plot in figure A1.6(a). In this case, the figure in terms of electrical power output is £660/kWe.

A1.11 Financial Evaluation
Finally, figures for installation and maintenance costs are entered into table 2.6(a) and the necessary calculations are undertaken to work out net revenue savings and total capital cost.

The final figures for the scheme evaluated for this example site are presented in table A1.6(a). With a total capital cost of £530,000 and net revenue savings of £119,000 per annum, the simple payback for the proposed scheme is 4.5 years. At a discount rate of 8% and taking the life of the installation as 15 years, the scheme has a net present value of £580,000.

Table A1.12(a) *Comparative CHP Scheme Financial Performance*

CHP Plant Size – kWe	Revenue Savings – £ pa	Total Capital Cost – £	Simple Payback – yrs	Net Present Value[1] – £
800	119,000	530,000	4.5	490,000
600	102,000	390,000	3.9	480,000
500	85,000	330,000	3.9	400,000

Notes

[1] Net present value calculations are based on a discount rate of 8% and an installation life (and hence investment period) of 15 years.

A1.12 Scheme Optimisation

For a site such as this, where base load never falls below 50% of the average load during 'major use hours', the optimum engine size from a financial perspective is likely to be closer to 'out of hours' average load than 'major use hours' average load. In addition, within the size range of IC engine based CHP plant considered here, cost per kWe of output is virtually constant and power generating efficiency increases only marginally with size.

For these reasons, the optimum financial return from a cogeneration scheme at this site would probably be achieved by a smaller installation than one sized on annual average electrical demand. Table A1.12(a) gives the results from preliminary appraisals executed for CHP installations sized at 800, 600 and 500 kWe, where 500 kWe is the base load for the site as indicated in figure A1.2(a).

Appendix 2

CONVERSION OF UNITS

The dual use of SI and Imperial (Inch-Pound) units in the UK and USA has led to a proliferation in the number of units that are in general use in engineering. The problem is particularly acute in the field of power generation. For this reason, conversion factors which will be of use to those undertaking work in the field of combined heat and power are given here.

APPENDIX 2: CONVERSION OF UNITS

A2.1 Universal Conversions

Pure unit conversions for various units in common use in the detailing of engine performance are given in table A2.1(a). These conversions are universal as they do not depend on the properties of the substance or process being measured.

A2.2 Exhaust Emissions

The various forms in which emissions data for engines can be presented have been discussed in section 5.9.1. Conversions between different forms are, unfortunately, not universal but depend upon the actual gas being quantified.

Formulae for converting between emissions units are given below.

Conversion between different Volume Based Units

$$ppm \ v/v \ = \ \frac{g/nm^3 \times 1000}{\rho_i}$$

where

ppm v/v = parts per million, volume by volume

g/nm^3 = grams per cubic metre at normalised conditions of 0°C and 101.3 kPa

Table A2.1(a) *Pure Unit Conversions*

Multiply	By	To Obtain
Area		
in²	0.0006452	m²
ft²	0.09290	m²
acre	4,047	ha
ha	10,000	m²
Energy		
Btu	0.001055	MJ
bhph	0.7457	kWh
kWh	3.600	MJ
therm	100,000	Btu
Force		
lbf	4.448	N
Fuel Consumption Rate		
MBtu/h	293.1	kW
Btu/min	0.01759	kW
MJ/h	0.2778	kW
Fuel Heat Value		
Btu/lb	0.002326	MJ/kg
therm/ft³	3725	MJ/m³
Length		
in	0.02540	m
ft	0.3048	m

Multiply	By	To Obtain
Mass		
lb	0.4536	kg
Mass Flow Rate		
lb/h	0.0001260	kg/s
Pressure		
psi	6.895	kPa
bar	100.0	kPa
in of mercury	3.386	kPa
in of water	0.2491	kPa
Power		
bph	0.7457	kW
Btu/h	0.0002931	kW
Specific Fuel Consumption		
lb/bhph	0.6083	kg/kWh
Btu/bhph	0.0003930	1/Efficiency
MJ/kWh	0.2778	1/Efficiency
Volume		
gallon (UK)	4.546	*l*
gallon (US)	3.785	*l*
l	0.001	m³
ft³	0.02832	m³
Volume Flow Rate		
gpm (UK)	0.07577	*l*/s
gpm (US)	0.06309	*l*/s

ρ_i = individual emission density in kg/nm³
  2.035 for nitrogen oxides
  1.249 for carbon monoxide
  1.964 for carbon dioxide
  0.725 for non-methane hydrocarbons

Adjustment for different Oxygen Content for a Volume Based Unit

$$U_{v\,@\,y\%\,O_2} = U_{v\,@\,x\%\,O_2} \times \frac{20.95 - y\%}{20.95 - x\%}$$

where

$U_{v\,@\,y\%\,O_2}$ = volume unit in either ppm v/v or g/nm³ at y% exhaust gas oxygen content

Conversion between Fuel and Work Based Units

$$g/GJ = \frac{g/kWh \times \eta}{0.36}$$

where

g/GJ = grams per gigajoule of fuel burned by the engine

g/kWh = grams per kilowatt hour of shaft work from the engine

η = engine brake thermal efficiency in per cent

Conversion between Volume and Fuel or Work Based Units

$$g/nm^3 = \frac{g/GJ \times \rho_o \times LHV}{(1 + AFR) \times 1000}$$

$$g/nm^3 = \frac{g/kWh \times \rho_o \times LHV \times \eta}{(1 + AFR) \times 360}$$

where

ρ_o = overall exhaust gas density in kg/nm³
  1.22 for natural gas firing
  1.29 for distillate oil firing

LHV = lower heat value of fuel in MJ/kg
  47 for natural gas
  43 for distillate oil

AFR = air to fuel ratio
  19.7 for natural gas firing *to give an exhaust oxygen content of 5%*
  18.8 for distillate oil firing *to give an exhaust oxygen content of 5%*

Appendix 3

LEGISLATION, STANDARDS AND CODES OF PRACTICE

A listing of the key documents relating to combined heat and power installations in the UK is given here.

APPENDIX 3: LEGISLATION, STANDARDS AND CODES OF PRACTICE

A3.1 Engines and Alternators

- Technical Memorandum Number 3, 'A Code of Practice for Designers, Installers, and Users of Generating Sets', The Technical Standards Committee of the Association of British Generating Set Manufacturers, Bisley, UK, 1985 (currently in revision).

- BS 7698: 1993; (ISO 8528: 1993) 'Reciprocating Internal Combustion Engine Driven Alternating Current Generating Sets', HMSO, UK.
  Part 1: Specification for application, ratings and performance.
  Part 2: Specification for engines.
  Part 3: Specification for alternating current generators for generating sets.
  Part 4: Specification for controlgear and switchgear.
  Part 5: Specification for generating sets.

- IM/17, 'Code of Practice for Natural Gas Fuelled Spark Ignition and Dual-Fuel Engines', British Gas, London, UK, February 1981.

- IM/24, 'Guidance Notes on the Installation of Industrial Gas Turbines, Associated Gas Compressors and Supplementary Firing Burners', British Gas, London, UK, June 1989.

A3.2 Gas Supplies and Gas Compressors

- Statutory Instrument 1984 No. 1358, 'The Gas Safety (Installation and Use) Regulations 1984', HMSO, UK.

- IM/16, 'Guidance Notes on the Installation of Gas Pipework, Boosters and Compressors in Customer's Premises', British Gas, London, UK, 1989.

- IM/21, 'Guidance Notes on The Gas Safety (Installation and Use) Regulations 1984', British Gas, London, UK, September 1985.

A3.3 Electrical Switchgear and Protection

- Statutory Instrument 1988 No. 1057, 'The Electricity Supply Regulations 1988', HMSO, UK.

- Engineering Recommendation G59/1, 'Recommendations for the Connection of Embedded Generation Plant to the Regional Electricity Company's Distribution System', Electricity Association, London, UK, 1991.

- Engineering Technical Report No. 113, 'Notes of Guidance for the Protection of Private Generating Sets up to 5MW for Operation in Parallel with Electricity Board's Distribution Networks', Electricity Association, London, UK, 1989.

A3.4 Environmental Impacts

- 'Secretary of State's Guidance - Compression ignition engines, 20-50 MW net rated thermal input', PG1/5(91), HMSO, UK, February 1991.

- 'Secretary of State's Guidance - Gas turbines, 20-50 MW net rated thermal input', PG1/4(91), HMSO, UK, February 1991.

- Chief Inspector's Guidance to Inspectors, Process Guidance Note IPR I/2, 'Combustion Processes - Gas Turbines', HMSO, UK, 1992.

- Chief Inspector's Guidance to Inspectors, Process Guidance Note IPR I/3, 'Combustion Processes - Compression Ignition Engines 50MW(th) and over', HMSO, UK, 1992.

Appendix 4

BIBLIOGRAPHY

APPENDIX 4: BIBLIOGRAPHY

Text References

1 Rogers G.F.C. and Mayhew Y.R. 1979, "Engineering Thermodynamics Work and Heat Transfer", Longman, London.
2 ASHRAE HVAC Applications Handbook, American Society of Heating, Refrigerating and Air Conditioning Engineers, Atlanta.

A4.1 Cogeneration

- 1992 ASHRAE Handbook, HVAC Systems and Equipment, Chapter 7, 'Cogeneration Systems', American Society of Heating, Refrigerating and Air-Conditioning Engineers, Atlanta, GA, USA.

- CADDET Analysis Series No. 9, 'Learning from Experiences with Gas-Turbine-Based CHP in Industry', Centre for the Analysis and Dissemination of Demonstrated Energy Technologies, Sittard, Netherlands, 1993.

- 'A Review of Cogeneration Equipment and Selected Installations in Europe', OPET, for the Commission of the European Communities, Directorate-General XVII for Energy, Institut Valencia de l'Energia, Valencia, Spain.

- Good Practice Guide 43, 'Introduction to Large-Scale Combined Heat and Power', Energy Technology Support Unit, Harwell, UK, 1992.

- Good Practice Guide 60, 'The Application of Combined Heat and Power in the UK Health Service, Energy Technology Support Unit, Harwell, UK, 1992.

- Mesko J.E., 'Economic Evaluation of Various Cogeneration Systems for a Large University Campus', COGEN-TURBO IGTI-Vol.6, American Society of Mechanical Engineers, Fairfield, NJ, USA, 1991.

- Ballantyne J.G., Holmes D. and Marshall J.B., 'Twelve years experience with a 23MWe gas turbine CHP plant', Publication 481, Institution of Diesel and Gas Turbine Engineers, Bedford, UK, August 1994.

- Ward C.R. 'Combined cycle CHP power station at Purfleet', Publication 479, Institution of Diesel and Gas Turbine Engineers, Bedford, UK, April 1994.

- Ward D.D., 'Cogeneration retrofit study at Guy's Hospital', Publication 482, Institution of Diesel and Gas Turbine Engineers, Bedford, UK, February 1994.

- Cooke R., 'Experience of operating CHP plant', Publication 485, Institution of Diesel and Gas Turbine Engineers, Bedford, UK, November 1994.

- Pickering S.J. and Cooper P.W., 'Operation of a diesel engine CHP plant at the University of Nottingham', Combined Heat and Power, Proceedings of the Institution of Mechanical Engineers, London, UK, October 1994.

- Tranberg E., 'Combined heat and power plant using four different biofuels', Combined Heat and Power, Proceedings of the Institution of Mechanical Engineers, London, UK, October 1994.

- Blunt C.N., 'CHP systems in paper mills', Combined Heat and Power, Proceedings of the Institution of Mechanical Engineers, London, UK, October 1994.

- Wainwright R, 'The gas turbine based combined heat and power station at the University of Birmingham', Combined Heat and Power, Proceedings of the Institution of Mechanical Engineers, London, UK, October 1994.

- Stanley R.B., 'The American I Cogeneration Plant – Lessons Learned from the Gilroy Foods Cogeneration Plant', PWR-Vol.11, Cogeneration and Combined Cycle Plants – Design, Interconnection, and Turbine Applications, American Society of Mechanical Engineers, New York, NY, USA, 1990.

- Multari P.L., 'Operational Experiences in Cogeneration Projects – From Project Conception to Full Scale Operation', Proceedings Gas Turbine Power Generation Commercial, Economic and Operational Challenges, Institute of Energy, London, UK, May 1995.

A4.2 Small Scale Combined Heat and Power

- CADDET Analysis Series No. 1, 'Learning from Experiences with Small Scale Cogeneration', Centre for the Analysis and Dissemination of Demonstrated Energy Technologies, Sittard, Netherlands.

- 'Small-scale cogeneration in non-residential buildings', OPET, For the Commission of the European Communities, Directorate-General XVII for Energy, Istituto Cooperativo per l'Innovazione, Rome, Italy.

- Good Practice Guide 1, 'Guidance Notes for the Implementation of Small Scale Packaged Combined Heat and Power', Energy Technology Support Unit, Harwell, UK, 1989.

- Good Practice Guide 3, 'Introduction to Small Scale Packaged Combined Heat and Power', Energy Technology Support Unit, Harwell, UK, 1990.

- Evans R.D., 'Environmental and Economic Implications of Small-Scale CHP', Energy and Environmental Paper No. 3, Energy Technology Support Unit, Harwell, UK, March 1990.

- Maurer H.F., 'Small scale combined heat & power plants in Thames Water Provinces', Publication 483, Institution of Diesel and Gas Turbine Engineers, Bedford, UK, March 1994.

- Andrews D., 'Installation Aspects For Spark Ignited CHP Systems', Independent Power Generation Conference 1992, Institution of Diesel and Gas Turbine Engineers, Bedford, UK.

- Kidwai M.A.S., 'Use of Landfill Gas for Small Scale Power Production', PWR-Vol.16, Cogeneration Power Plants: Combined Cycle Design, Operation, Control and Unit Auxiliaries, American Society of Mechanical Engineers, New York, NY, USA, 1991.

A4.3 Reciprocating IC Engines

- Nylund I. and Rosgren C., 'The latest achievements in gas diesel technology and the experience from some power plant applications', Publication 476, Institution of Diesel and Gas Turbine Engineers, Bedford, UK, October 1993.

- Masters K.J., 'Lubricating Oil Analysis – What is it all about?, Publication 489, Institution of Diesel and Gas Turbine Engineers, Bedford, UK, 1995.

- Ruck N. and Downing J.A., 'Lubrication aspects of CHP systems, Combined Heat and Power', Proceedings of the Institution of Mechanical Engineers, London, UK, October 1994.

- 1992 ASHRAE Handbook, HVAC Systems and Equipment, Chapter 41, 'Engine and Turbine Drives', American Society of Heating, Refrigerating and Air-Conditioning Engineers, Atlanta, GA, USA.

- Marine Engineering Practice Volume 1, Part 2, 'Prime Movers for Generation of Electricity (B) Medium Speed Diesel Generating Sets', Institute of Marine Engineers, London, UK, 1974.

- Kiess R., 'Diesel Engine Development', Independent Power Generation Conference 1992, Institution of Diesel and Gas Turbine Engineers, Bedford, UK.

A4.4 Gas Turbines

- Ediss B.G., 'Steam injected gas turbine cycle', Publication 467, Institution of Diesel and Gas Turbine Engineers, Bedford, UK, November 1991.

- Meher-Homji C.B., 'Gas Turbine Axial Compressor Fouling – A Unified Treatment of its Effects, Detection, and Control', COGEN-TURBO IGTI-Vol.6, American Society of Mechanical Engineers, Fairfield, NJ, USA, 1990.

- Seddigh F. and Saravamuttoo H.I.H., 'A Proposed Method for Assessing the Susceptibility of Axial Compressors to Fouling', Transactions Vol.113, American Society of Mechanical Engineers, Fairfield, NJ, USA, 1991.

- Diakunchak I.S., 'Performance Deterioration in Industrial Gas Turbines', Transactions Vol.114, American Society of Mechanical Engineers, Fairfield, NJ, USA, 1992.

- 1992 ASHRAE Handbook, HVAC Systems and Equipment, Chapter 41, 'Engine and Turbine Drives', American Society of Heating, Refrigerating and Air-Conditioning Engineers, Atlanta, GA, USA.

A4.5 Steam Turbines

- 1992 ASHRAE Handbook, HVAC Systems and Equipment, Chapter 41, 'Engine and Turbine Drives', American Society of

Heating, Refrigerating and Air-Conditioning Engineers, Atlanta, GA, USA.

- Marine Engineering Practice Volume 1, Part 2, 'Prime Movers for Generation of Electricity (A) Steam Turbines', Institute of Marine Engineers, London, UK, 1973.

- Engelke W., Bergmann D. and Termuehlen H., 'Steam Turbines for Combined Cycle Power Plants', PWR-Vol.11, Cogeneration and Combined Cycle Plants – Design, Interconnection, and Turbine Applications, American Society of Mechanical Engineers, New York, NY, USA, 1990.

A4.6 Boilers and Heat Recovery Equipment
- Lister D., 'Modern boilers for CHP applications', Combined Heat and Power, Proceedings of the Institution of Mechanical Engineers, London, UK, October 1994.

A4.7 Electrical Switchgear and Protection
- Sheldon R. 'Parallel Operation with the REC's Distribution System', Independent Power Generation Conference 1992, Institution of Diesel and Gas Turbine Engineers, Bedford, UK.

- Rogers W.J.S., 'Connection of Generators for Parallel Operation with a Network Supplying the General Public', Proceedings Protecting Electrical Networks and Quality of Supply in a De-Regulated Industry, Report 95-0002, ERA Technology Ltd., Leatherhead, UK, February 1995.

- Woodworth M.H., 'The Integration of Embedded Generation onto the Public Electricity Supplier's Networks', Proceedings Protecting Electrical Networks and Quality of Supply in a De-Regulated Industry, Report 95-0002, ERA Technology Ltd., Leatherhead, UK, February 1995.

- Shaw T.O.R., 'Case Studies of Operating Considerations Omitted from Design of Local Electrical Networks', Proceedings Protecting Electrical Networks and Quality of Supply in a De-Regulated Industry, Report 95-0002, ERA Technology Ltd., Leatherhead, UK, February 1995.

- Russell J.C., 'Protection Relays for Mains Failure with respect to G59', Proceedings Protecting Electrical Networks and Quality of Supply in a De-Regulated Industry, Report 95-0002, ERA Technology Ltd., Leatherhead, UK, February 1995.

A4.8 Automatic Control and Monitoring
- Allen T. et al, 'Energy Management Software for a Combined Cycle Cogeneration Facility', PWR-Vol.16, Cogeneration Power Plants: Combined Cycle Design, Operation, Control and Unit Auxiliaries, American Society of Mechanical Engineers, New York, NY, USA, 1991.

- Gant G.C., Nevard R.J. and Poppe R.H., 'Fault Detection and Diagnosis in Diesel Engines', Independent Power Generation Conference 1992, Institution of Diesel and Gas Turbine Engineers, Bedford, UK.

- Bower I., McCune J. and Starkey D., 'Updating Cogeneration Control Systems for Improved Reliability and Operability', PWR-Vol.16, Cogeneration Power Plants: Combined Cycle Design, Operation, Control and Unit Auxiliaries, American Society of Mechanical Engineers, New York, NY, USA, 1991.

A4.9 Environmental Impacts
- 1993 ASHRAE Handbook, Fundamentals, Chapter 7, 'Sound and Vibration', American Society of Heating, Refrigerating and Air-Conditioning Engineers, Atlanta, GA, USA.

- 1995 ASHRAE Handbook, HVAC Applications, Chapter 43, 'Sound and Vibration Control', American Society of Heating, Refrigerating and Air-Conditioning Engineers, Atlanta, GA, USA.

- DiCola F.E. and Winge D.E., 'Correction of Noise Complaints at a Cogeneration Plant', PWR-Vol.16, Cogeneration Power Plants: Combined Cycle Design, Operation, Control and Unit Auxiliaries, American Society of Mechanical Engineers, New York, NY, USA, 1991.

INDEX